Sir William de Wiveleslie Abney

Colour Vision

Being the Tyndall Lectures delivered in 1894 at the Royal Institution

Sir William de Wiveleslie Abney

Colour Vision
Being the Tyndall Lectures delivered in 1894 at the Royal Institution

ISBN/EAN: 9783337070489

Printed in Europe, USA, Canada, Australia, Japan

Cover: Foto ©berggeist007 / pixelio.de

More available books at **www.hansebooks.com**

COLOUR VISION

TYPES OF COLOUR VISION.

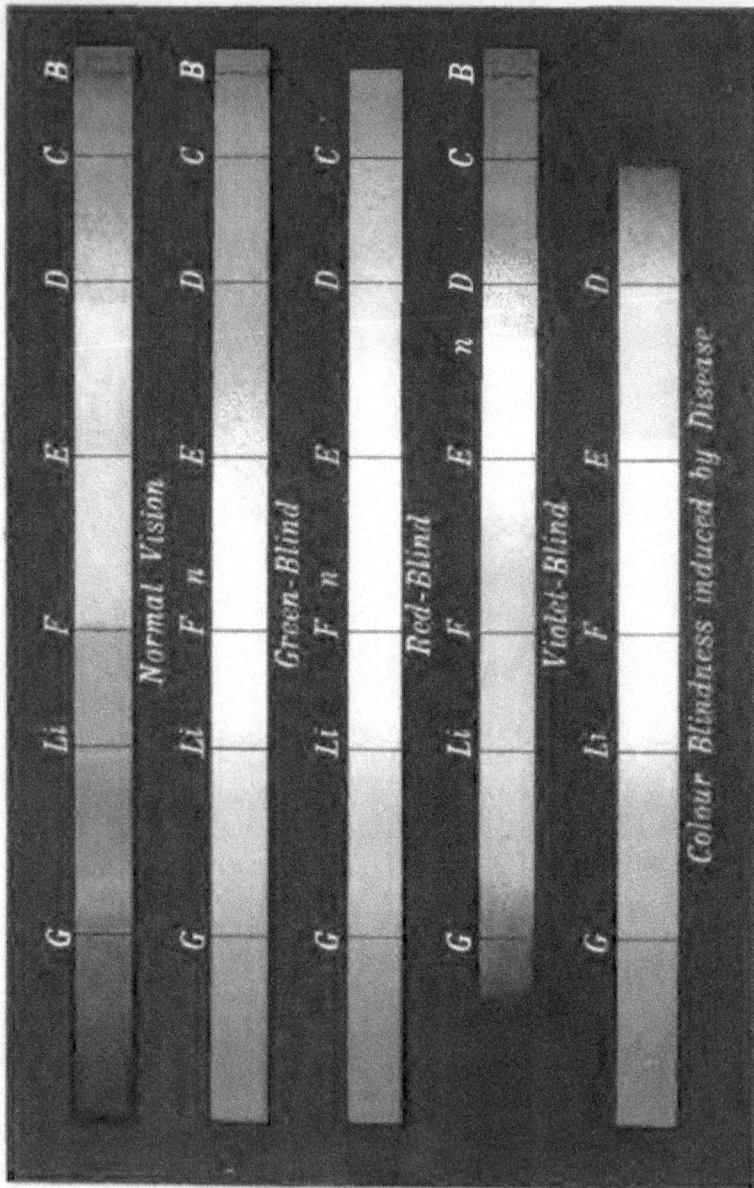

No. 1. Normal Vision

No. 2. Green-Blind

No. 3. Red-Blind

No. 4. Violet-Blind

No. 5. Colour Blindness induced by Disease

W. DE W. A.-, DEL.

W. GRIGGS, LITH.

COLOUR VISION

BEING

(24)

THE TYNDALL LECTURES

DELIVERED IN 1894

AT

THE ROYAL INSTITUTION

BY

(Sir)

CAPT. W. DE W. ABNEY, C.B., D.C.L., F.R.S.,

LATE ROYAL ENGINEERS

WITH COLOURED PLATE AND NUMEROUS DIAGRAMS

LONDON

SAMPSON LOW, MARSTON & COMPANY

LIMITED

St. Dunstan's House

FETTER LANE, FLEET STREET, E.C.

1895

CONTENTS.

CHAPTER VIII.

CHAPTER IX.

CHAPTER X.

CHAPTER XI.

CHAPTER XII.

CHAPTER XIII.

CHAPTER XIV.

CHAPTER XV.

PREFACE.

THE writer had for some years past, in conjunction with General Festing, and recently as Secretary and Member of the Colour Vision Committee of the Royal Society, carried out a series of investigations on colour vision, and selected that subject when he was invited, in 1894, to deliver the Tyndall Lectures at the Royal Institution.

The brief time allotted for these lectures—an hour on three successive Saturday afternoons—restricted the discussion of some aspects of the question, and confined its treatment in the main to those features most readily explicable by the physicist, and to bringing into notice the latest results which had been obtained from physical experiments. How far the

writer has succeeded in the task which he then out-
lined it is for the reader to determine.

There was no intention in the first instance to
publish these lectures. After their delivery, many
persons expressed a desire that the information they
contained should be rendered accessible to such as
were interested in the theory of colour vision, and
in deference to that desire the lecture-notes have
been re-cast in book form. For the reader's convenience the matter is now divided into chapters
instead of into lectures, and a few additions have
been made in the text to explain some of the experimental work to those who have not facilities for
its repetition.

The writer has to acknowledge several debts of
gratitude. First, to Mr. E. Nettleship, for his kindness
in looking over the proofs, and making valuable sug-
gestions whilst the work was passing through the
press; and also, as will be seen throughout its pages,
for many of the interesting cases of defective colour
perception which have been examined by the some-

what novel methods described. Next, the writer's
gratitude is due to Professor M. Foster for the permission he has given to use his admirable description
of the Hering theory; and, lastly, to the Royal Society
for the permission it accorded to use various diagrams
which have served as illustrations to papers which
have appeared in its " Philosophical Transactions"
and " Proceedings."

COLOUR VISION.

CHAPTER I.

I MUST commence this course by saying that I feel the honour that has been done me in asking me to undertake it, connected as it is with the name of Tyndall, whose recent removal from our midst has been deplored by all lovers of science, and by none more than by those who have had the privilege of listening to him at this Institution. It is my duty to speak on some subject of physics, and the subject I have chosen is Colour Vision. I hope it will not be considered inappropriate, since it was Thomas Young, the physicist, whose connection with this Institution is well known, who first propounded a really philosophical theory of the subject. Interesting as it may be to trace how old theories have failed and

B

new ones have started, I feel that for those who, like
myself, have but little time at command in which to
address you, the historical side of this question must
of necessity be treated incompletely.

Colour vision is a subject which enters into the
domains both of physics and physiology, and it is
thus difficult for any one individual to treat of it
exhaustively unless he be a Helmholtz, who was as
distinguished in the one branch of science as he was
in the other.   I am not a physiologist, and at the
most can only pretend to an elementary knowledge
of the physiology of the eye, but I trust it is
sufficient to prevent myself from falling into any
grievous error.   I shall try and show you, however,
that the subject is capable of being made subordinate
to physical methods of examination.   I must necessarily commence by a very brief description of those
parts of the eye in which it is supposed the seat of
vision lies, but in terms which are not too technical.
As to the mere optical properties of the eye I shall say
but little, for they are not necessary for my purpose,
although more particularly adapted to mathematical
treatment than the other properties I have to describe.

The eye may be diagrammatically represented as
in the figure which is supposed to be a horizontal

section of it, the figure being reproduced from Professor Michael Foster's Physiology.

As far as the perception of colour is concerned, the principal part of the eye which is not distinctly optical —*i.e.* for the production of images — is the retina, and this it will be seen is in reality an outcrop of the brain, the connection between the two being the optic nerve. Owing to this connection, it is not easy to determine where the seat of colour perception is located; but for the purpose of physical investigation this is not of first-rate importance, nor

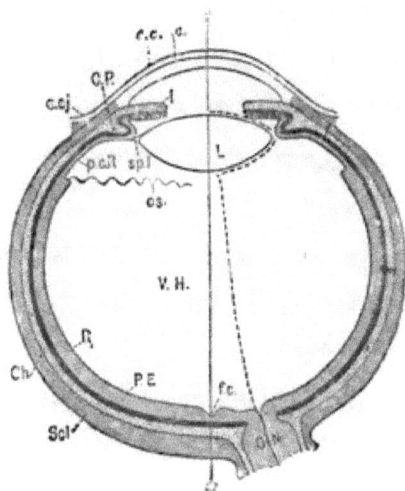

FIG. 1.

Scl is the sclerotic coat. *Ch* the choroid coat, with *CP* the ciliary process. *I* is the body of the Iris. *R* is the retina or inner wall. *PE* the pigment epithelium or outer wall. *L* the lens held by the suspensory ligament *sp.l.* *VH* is the vitreous humour. *ON* the optic nerve. *ox* is the optic axis, in this case made to pass through the fovea centralis, *f.c.*

does it affect the discussion of rival theories except in a minor degree. There are other subsidiary adjuncts in the eye to which, however, I must call attention, as they have a distinct bearing on the apparent intensity of some colours and of the hue that mixtures of others are perceived. The first is what is called the "macula lutea," or yellow spot, a spot which it may be assumed exists in every eye. It is horizontally oval in form, and is situated in the very centre of the retina, embracing some 6° to 8° in angular measure. It has a brownish or yellowish tint, and the retina at this part is slightly depressed, being bounded by a slightly raised rim. In the centre of this area the retina becomes very thin, having a depression about $\frac{1}{100}$ of an inch or ·3 millimetres in diameter, which is named the "fovea centralis," where it is said that vision is the most acute. This statement can be well credited when we come to consider where the seat of the stimulation of sensation lies. The colour which tints the yellow spot is strongest at the crater-like rim, and fades away centrally and peripherally, and is said to be wholly absent in the fovea centralis.

As the colour of this spot is yellow or brown in the living eye (and that it is probably brown the absorption indicates), it follows that white light passing through it must be deprived of some of its components, though in differing degrees. If the seat of sensation is at the outer layer of the retina, as we shall shortly see must be the case, it will further be seen that when light of any colour which the brown pigment will absorb more or less completely falls on different parts of the oval area, the absorption must vary at each part, and the intensity of the perceived light will be least at the rim and increase centrally and peripherally. As the centre of the yellow spot or fovea is coincident approximately with the point where the axis of the eye cuts the retina, the image of an evenly illuminated object, when looked at directly, must fall on the yellow spot. If, therefore, a patch of such light, the image of which more than covers the spot, be observed, it ought to exhibit a varying brightness of colour corresponding to the strength of the colouring matter which exists at the different parts. This it but rarely does, for habit and constant interpretation of what should be seen prevents the mind from distinguishing these variations; but if the colour brightness, as perceived by the different parts, be submitted to measurement by proper means, the variations in brightness of the image can be readily recognised. A very common method of exhibiting the presence of the pigment is to look at a bright white cloud through a layer of chrome alum. Chrome alum transmits red and blue - green rays. Now as the spectrum-blue rays are those which the pigment will absorb, it follows that the colour of the solution should appear ruddy to the central part of the eye, but on the rest of the retina it should appear of its ordinary purplish colour. At a first glance, and before the eye has become fatigued, this is the case, but the phenomenon soon disappears. Another way of forming an idea as to what the yellow spot absorbs is to throw a feeble spectrum on a white surface and cause the eye to travel along it. If the spectrum be viewed so that it does not occupy more than about 40° of the retina, the movement of the eye will show a dark band travelling along the green, blue, and violet regions as the image of these parts of the spectrum fall on the yellow spot, and their apparent brightness will increase as they fall outside the absorbing area. This proves that an absorption takes place in this area.

The retina consists essentially of an inner and outer wall, enclosing matter which is similar to the grey matter of the brain. On the inner wall are the vessels which are connected with the optic nerve. The outer wall is epithelium coloured with a pigment, and it is here that the visual impulses begin, although the rays of light giving rise to them have to pass through the thickness of the retina before so doing. It has already been stated that the light has to pass through the thickness of the yellow spot before a visual sensation is felt in the centre of the field, and the experiments just given offer a fair proof of the truth of the assertion, but there is still another which is perhaps more conclusive. Suppose we have a hollow reflecting ball, as shown in Fig. 2, and through an orifice A we project a

Fig. 2.

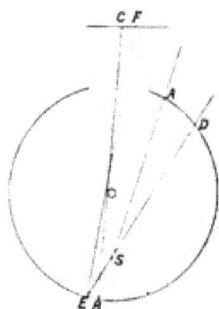

beam of light to B, which meets an obstruction, S, in its path, then A B would be reflected from B to C on a screen C F, and the obstruction S would be marked at C. If another beam from D was directed so as to meet the same obstruction, its presence would be marked at F. Knowing the distance of the centre O of the hollow sphere from F C and its diameter, and measuring the distance between F and C and their respective distances from the axis of the sphere, the distances S B and S E can be calculated. This method is applied in the formation of what are known as Purkinje's figures. The simplest case is where a beam of light is directed through the sclerotic and transmitted through the lens. Images of the retinal

vessels are distinguished as at S, and it is found
that they cast shadows, which are seen as dark lines
in the glare of the field of vision. The sensation
of light must therefore come from behind these
vessels, and calculation shows that the seat of the
sensation is close to the pigmented inner wall of the
retina.

Lying here is a layer of what are known as rods
and cones, which have a connection, either actual or
functional, with the optic fibres which largely compose
the inner wall of the retina, and are connected with
the optic nerve. In the yellow spot the cones are
much more numerous than the rods, but in the peripheral part the reverse is the case. In the fovea the
rods appear to be altogether absent. The total number
of cones in the eye has been calculated to be about
3,000,000, of which about 7,000 are in the small fovea.
The number of cones will give an idea of their dimensions. This detail has been entered into as it has been
supposed that these rods and cones are all-important
in translating light-waves into visual impulses. The
inner wall of the retina of most human eyes, as has
been mentioned, is stained with a black pigment,
fuscin, though in albinos it is absent. What its particular use may be is still unknown, for its change by

light is so slow that it can scarcely be the cause of vision. In the outer parts of the rods is, however, diffused a substance highly sensitive to light, called the "visual purple," from its colour, and a theory founded on chemical action. produced by a change in this substance, has been promulgated. Fascinating, however, as such a theory must be, it lacks confirmation. The fact that the cones do not contain it, and that in the fovea are cones alone, renders it difficult to reconcile the theory with the fact that this part of the retina possesses, we are told, the greatest acuteness of sensation as regards light and colour.

The eyes of most vertebrate animals, it may be remarked, have this visual purple, but in those of the bat, owl, hen, and some others the colouring matter seems to be absent. Visual purple is an interesting substance, however, and as it is found in the eye it probably exercises some useful function, though what that function may be is at present unknown. That images of objects can be formed on the retina, owing to the bleaching of this substance, has been proved by experiment. The purple is first changed to a yellow colour, and then passes into white. These "optograms," as they are called, can be fixed in an excised eye if the retina be detached, and then be treated with a weak solution of alum.

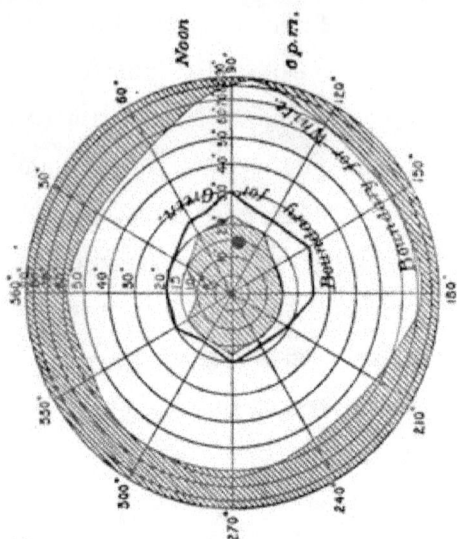

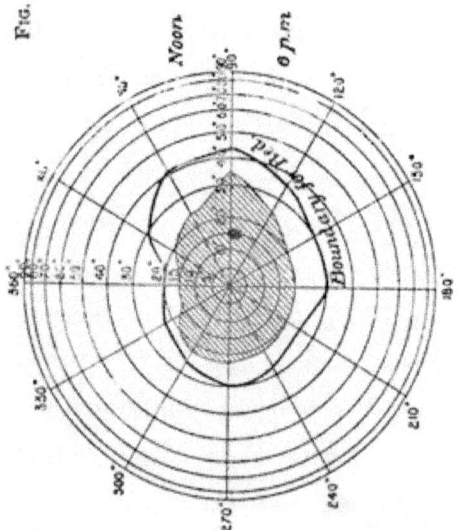

Many persons are not aware of the extent of the field of view which the eye embraces. Vertically it takes in about 100°, whilst horizontally it will take in some 145°, more or less. The field is smaller on the nasal than on the temporal side. When both eyes are used, the combined field of view is larger horizontally, being about 180°. The field of view which is common to both eyes is roughly a circle of about 90°. There

is, however, a marked difference in the distinctness
with which objects are perceived in the different parts
of field of view. On the fovea centralis two dots placed
so as to subtend an angle of 60″ will be perceived as
double. That is to say, if a piece of paper, on which
are two dots $\frac{1}{30}$ of an inch apart, be placed 10 feet
away from the observer, these dots will be seen as
separated, whilst dots (in this case they should be
black and of good dimensions) placed half-an-inch
apart would still appear as one if viewed at the same
distance near the periphery of the retina. In the
yellow spot the distance apart of the cones is such
that they subtend about the same angle as the dots
when they are seen separate, viz., about 60″; that is,
they are about $\frac{16}{100000}$ of an inch apart, and hence
may have something to say to the limit of separation.
The field for the perception of colour is different to
that for light.

The diagrams (Fig. 3) will show the fields in a
satisfactory manner. The concentric circles are supposed to be circles lying on the retina corresponding
to parallels of latitude on a globe, and are not, therefore, equi-distant when seen in projection. To make
these circles it must be imagined that we have a bowl,
in the middle of which is a thin rod standing upright

and passing through the centre, and another rod attached to it at the centre of the sphere of exactly the length of the radius. If this last arm be opened to make an angle of 5° with the fixed rod, and be twisted round like the leg of a compass against the bowl, it will make a circle, the projection of which will give the innermost circle of the diagram; if opened to 10° it will give the next circle, and so on for every subsequent 10°. The lines passing through the centre are 30° from one another, the line stretching from 360° to 180° being a line supposed to be vertical. By means of an instrument called the perimeter, the field of vision for each eye can be measured. With its aid any small object can be made to fall on any part of the retina by directing the axis of the eye to a fixed point and moving the object along one of the diameters. Suppose we wish to ascertain the field for a white object, a small white disc is moved, say, along the horizontal line, and the angles at which the retina just no longer sees it are noted. This gives two points in the field, and they are plotted on the chart—in Fig. 3 one touches the outside circle, and the other is at an angle of about 65°. The field of vision is next tested along another line, say 300° to 120°, and other points noted and marked on the chart. When

the whole circle has been examined, the various points
are joined together, and we have the boundary of
vision for a white object. The boundaries of the *colour*
perception for (say) small red and green discs are found
in the same way. The former is depicted in the left-
hand chart and gives the field for the right eye, and
the latter with that for white in the right-hand chart
for the same eye. It will be noticed that two boundaries are given, one taken at mid-day and the other
at 6 p.m. The brighter the colour, the larger is the
boundary in both cases, showing that the field of colour
vision varies according to the illumination. Now it is
difficult from this method of experimenting to determine
whether the fields for different colours are the same
or differ in extent, as we have no information as to
whether the colours themselves which were used were
physiologically equal. The only way by which this can
be satisfactorily determined is by using spectrum colours
each of known brightness and area. (Some preliminary
experiments made by myself regarding the colour fields
will be found in the appendix, and will be referred to
later.) It must not be thought that the various colour
boundaries mark the limit at which *light* is perceived,
but only the limit at which colour is seen ; outside the
boundaries the objects appear of a nondescript colour,

to which we shall by-and-by call attention. The yellow spot lies within the circle of 5°, and the blind spot on which no sensation of light is stimulated is shown by the black dot about 15° away from the centre.

I have only attempted to sketch, in unphysiological language, the primary apparatus with which our experiments in colour have perforce to be made.

CHAPTER II.

It will be seen, then, that in measuring colour or light several circumstances have to be taken into account. These are not simple, and require differentiating one from another before the results of colour measures can be finally laid down as correct, or as being held to be applicable to all cases.

We must naturally ask, what is colour? The answer I should like to pass over entirely. It can only be described as a sensation, just as we should describe touch as a sensation. It has, however, one advantage over most sensations, in that it is a sensation which can be submitted to empyric measurement. The question whether certain phenomena, such as the colours produced by simultaneous contrast, are subjective or real, does not require answering for the purpose that we have in view, but the results recorded may probably help to throw light on it. Colour is an impression caused by the stimulation in the eye

of some apparatus, that lies near the outer wall of
the retina, the effect of the stimulation being con-
veyed by the optic nerve to the brain.   If this
apparatus be complicated by being made up of
distinct parts, each of which transmits its own kind
of impression to the brain, it is not only quite
possible, but more than probable, that when one part
is absent or injured the particular impression for which
it is responsible will be lacking, and that the sum of
the impressions due to the remainder will be unlike
that perceived when they are all working together.

In every investigation, whether it be in physical or
in any other branch of science, it is better to work
up from the simple to the more complicated; and
acting on this plan, it is better to commence experimenting with simple rather than with complex colours,
though they may apparently produce precisely the
same sensations.   I shall, with this in view, devote
most of the remaining part of this chapter to some
necessary experiments with simple colours.   The simple
colours are those of the spectrum, and are the result
of motion in the ether, which pervades all space.   The
motion is in the form of undulations or waves, and
each colour is due to a series of these waves, which
have a definite length.   Thus, 6562 ten-millionths of

a millimetre produces to most of us a red colour in the spectrum (see Plate I.), occupying the position indicated by a black line known as the C line in the solar spectrum.

A table of wave-lengths of certain lines in the solar spectrum is given below :—

TABLE OF WAVE-LENGTHS IN TEN-MILLIONTHS OF A MILLIMETRE.

B, deep red	. .	6866	b, green . . .	5183
Lithium, cherry red .		6705	F, bluish green .	4861
C, red	. . .	6562	Lithium, blue .	4603
D, orange	. .	5892	G, violet. .	4307
E, green .	. .	5269	H, extreme violet	3968

The rays in the different parts of the spectrum being due to these simple vibratory motions, cannot be decomposed further. And it makes no matter whether we *see* them as different colours or not, they will always issue at the same angle from the same prism (if the prism be used to form the spectrum), when it is turned to the same angle to the incident light. Milestones are useful along a road to tell us where we are in reference to some central place, and these black lines in the spectrum serve the same end. But they have the advantage over the milestone, for whilst the last will tell us how far we are from, say, York or London, the former tell us our distance from

C

a zero point. We thus have a scale of light of different wave-lengths laid down for us, which we can apply to the study of the sensations stimulated in the eye, and so have the means of instituting a comparison between the colour vision of different eyes. A mixed or composite colour is in a different category, however, to the simple colour, as you will see directly. It is one which may be formed by any number of rays of different wave-lengths falling on the eye. What these rays are we can only tell by analysing the light and referring them to the spectrum.

The instrument before you is one which I have used before in this theatre; but as the major part of my experiments have been carried out with it, in case those who are present may not be acquainted with it, it will be necessary to describe it very briefly. The general arrangement of the apparatus is given in the accompanying diagram, Fig. 4.

R R are rays coming from the source of light, be it sun light or the electric light, and an image of the one or the other is formed by a lens L_1 on the slit S_1 of the collimator C. The parallel rays produced by the lens L_2 are partially refracted and partially reflected. The former pass through the prisms P_1, P_2, and are focussed to form a spectrum at D by a lens L_3. D is a

movable screen in which is an aperture S_2, the width
of which can be varied as desired. The rays are again

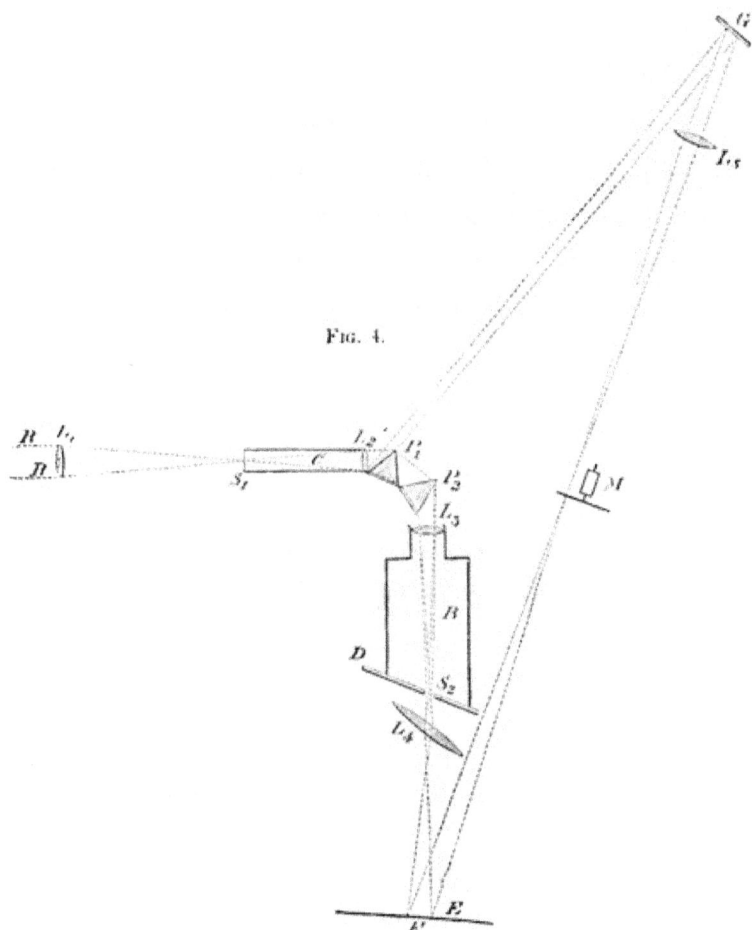

Fig. 4.

collected by a lens L_4, and form a white image of the
surface of the last prism on the screen E. If the light
passing through S_2 is alone used, the image at E is

formed of practically mono-chromatic light. Part of
the rays falling on P_1 are, as just said, reflected, but
as it and the refracted part are portions of the light
passing through the slit S_1, they both must vary proportionally. If then we use the reflected portion as
a comparison light to the spectrum colours, the relative intensities of the two, though they may vary
intrinsically, will remain the same. The rays reflected
from P, fall on G, a silver or glass mirror, and, by
means of another lens L_3, also can be caused to form
a white patch on the screen E, alongside the patch
of colour. At M, or anywhere in the path of the
beams, an electro-motor driving a sector with apertures which can be opened or closed whilst rotating,
is placed, and the illumination of either beam can be
altered at will. To obtain a large spectrum on the
screen E, all that is necessary is to interpose a lens
of fairly short focus in front of L_4, when a spectrum
of great purity and brightness can be formed.

If it be required to measure the width of the slits S_2
(which we shall see further on is often necessary), a
small lens of short focal.length placed behind L_4 and
near the slit will cast a magnified image on E, and by
means of a scale placed there, the widths of each slit,
if there are more than one, can be read off on the

scale by bringing them successively into the same colour.

Originally the comparison light was a candle, and it answered its purpose fairly well, and for obtaining absolute measures is convenient at the present time. Fig. 5 will show its arrangement, but as both the candle and the electric light may vary independently of each other, it will be seen that for merely the comparison of the different spectrum colours, the previous arrangement is the better. In both cases the two beams — the direct and the comparison — may be made to cast shadows

FIG. 5.

by placing a rod in their path, the shadow cast
by one light is then illuminated by the other light.
By moving the rod towards or from the screen the
shadows can be brought side by side.

With this instrument it is easy to demonstrate that
a mixed colour may be mistaken for a simple colour
of the spectrum. In a glass cell with parallel sides
is a solution of potassium bichromate, which, to myself
and probably most of you, has a beautiful orange colour.
The spectrum of white light is now on the screen, and
if this orange liquid is placed in the path of the white
light before it reaches the prisms, all the violet, blue,
and most of the green is cut off, leaving some green-
yellow, orange and red only on the screen. That
these form the orange colour of the bichromate is
readily shown by removing the auxiliary lens. The
spectrum, which has its focus at D, is now recombined
into a patch of light, which is at once seen to be the
colour of the solution.

The colour of the bichromate is therefore a complex
or mixed colour according to our definition, for it is
made up of a large number of simple colours. What I
desire to show, however, is that this complex colour can
be mistaken by the eye for a simple colour. First, let
us interpose the cell with the bichromate in the path

of the *reflected* beam, and throw the patch of light
formed by it on a white surface A (Fig. 6), alongside
the patch of light B formed by the spectrum. Next
let us pass a single aperture (Fig. 7), which can be
opened and closed by a screw arrangement, through
the spectrum. By careful movement we at length
come to an orange ray, which is spread out by the
apparatus to form a patch on B, that to the majority
(and the word majority is
used with intention) of people
exactly matches the colour
of the bichromate. Thus we
have a proof that, as far as
the eye is concerned, the
simple and the complex

FIG. 6.

colours are identical. This illustration of the want of
power of the eye to analyse colour might be repeated as
often as we like. We may pass coloured wools, for instance, through the length of the spectrum and show
that they have the property of appearing bright in,
and therefore of reflecting, some colours and of almost
disappearing in others—a sure indication that these
colours are mixed colours as they are made up of the
rays which are reflected. Yet when viewed in white
light they can in many cases be matched with simple

colours in the way we matched the colour of the bichromate solution. This tells us that there is something which requires investigating as to the constitution of the perceiving apparatus, and points to the probability that it is less complicated than it would be were it able to differentiate, without the aid of the spectrum, between simple and complex colours. If the eye had a separate apparatus—and when I say apparatus I use the word for want of a better—for taking up the impression of every simple colour, it might well be assumed that a differentiation must take place.

There is one class of colours, it must be remembered, which can never be mistaken for simple colours. I refer to the purples—mixtures of red and blue—for there are no spectrum colours which unmixed can possibly match them. All other colours, as no doubt will soon be apparent, can be referred to some one spectrum colour, either in its pure state or else mixed with some variable quantity of white light. We are all familiar with the fact that there are three primary colours, and we are naturally led to consider these in the light of the experiments just made. As good a definition as any other of a primary colour is that it is a colour which cannot be formed by the mixture of any two or more colours. The original investigators in colour phenomena were the artists, and they found that neither red, nor yellow, nor blue could be formed by any mixture of pigments on their palette, but that all other colours could be made by a mixture of two or more of these three. Hence to these three were given the name of primary colours. When, however, the physicist began to work with the simple colours of the spectrum, it was speedily found that, at all events, the yellow was not a primary colour, as it could be formed by a mixture of green and red, whilst a green could not be formed by a mixture of any other two colours. This we can prove with our apparatus.

Three apertures, all of which can be opened or closed as required (see Fig. 7), are placed in the spectrum, one in the red, one in the green, and one in the violet. The last we shall not require at present, so it is entirely closed; but we vary the width of the other two. We find that with a little red added to a bright green, a yellow green is produced; with more red added we have yellow; with still more red, an orange. The relative brightness of

FIG. 7.

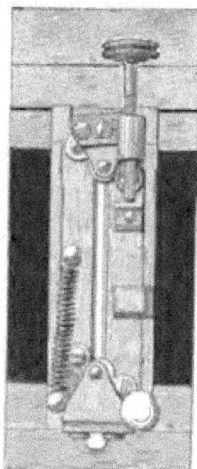

the two colours mixed together can be shown by removing the lens which recombines the spectrum to form
the patch of light. Each colour issues through its slit
and forms its own patch on a white screen which, for

Fig. 8.

the purpose, we make rather larger
than usual. The two patches overlap
in the middle (Fig. 8), and the pure
colours are seen one on each side of
the mixed colours.

Now, placing one slit in the yellow
and another in the blue of the spectrum, we find that
whatever width of slit we take, no green is produced,
but that, in fact, a yellowish or a bluish white results,
and that when the two slits are properly adjusted, a
pure white is produced. Evidently since none of the
intermediate spectrum colours between the blue and the
yellow can be made by their mixture, certainly green
cannot. Hence, with pure colours a green and not a
yellow is one of the primaries.

Further investigation on these lines has placed the
violet of the spectrum as a primary rather than the blue,
but this is still a matter of debate. Suffice it to say
that a red and a green in the spectrum are really two of
the primary colours, and most probably the violet the
third. Experiment shows that there is no other

primary colour in the strict sense of the word. We thus arrive at the fact that, except the primary colours themselves, every colour in nature may be made by a mixture of two or three of these primaries.

Just a word of explanation as to why, with pigments, the primary colours appear to be red, yellow, and blue, and not red, green, and blue. The colour of a pigment, it must be recollected, is a complex one. If we analyse a yellow—a yellow glass will be just as good an example as anything else—we find it is made up of green, yellow, orange. and red. A blue is made up of blue and green. If a yellow is placed behind a blue glass, and we look at a white surface through them, the only light that can get through the glass is the green. If the light, coming through each glass *separately*, falls on the same spot on a white surface, it will be either colourless or bluish white, or yellowish white, whichever colour preponderates. As the light reflected from mixed pigments is made up principally by the light coming through the different particles, first coming through one and then through another, and only partially by mixed lights, it will be gathered why the primary colour, when deduced from experiments with pigments, was yellow, and not green.

With the spectrum colours there is this fact to

remember, that though all intermediate colours between the pairs of primaries can be formed by their mixture, yet in some cases the resulting colours are *slightly* diluted with white, and that they thus appear less saturated than the spectrum colours themselves. The reason for this we shall be able to account for when we consider the colour sensations themselves.

When making matches to simple or other colours by the method of mixtures, we have to be careful of the conditions under which we experiment. This can be shown by a very simple experiment. I will make a match on B with the white light, which is thrown on the surface A (Fig. 6), by mixing the red, green, and violet that pass through the three adjustable apertures or slits already described. The apertures are altered till the match appears to myself perfect. From an appeal made to those of the audience who are at least 25 feet away from the patches of light, as to the correctness of the match, I gather that the match is to them imperfect. The mixed colours appear to them to give a pinkish white. The reason of this defect in the match is due to the fact that, as the lecturer is viewing the two square patches of 2 in. side from a distance of 2 ft. 6 in., their images on his retina extend beyond the boundary of the yellow spot, whilst the audience receives the whole

of the image on that portion of the retina which is
completely covered by it. To the lecturer only part of
the blue and green is absorbed by the yellow spot, and
the part of the retina outside it on which the image
falls receives and records the full intensity of these
colours. To the audience the full amount of absorption
takes place, with the result that the patch of mixed
colours must appear too red when it is correct to the
lecturer. In this case habit makes the eye take an
average of the different intensities which must exist at
the various parts of the image. We can, however, cause
a perfect agreement between all parties if the experi-
menter views the surfaces in a mirror placed some 12
feet away and then makes the match, for he is viewing
the patches from what is practically a distance of 24
feet. If after making the match without the aid of the
mirror the lecturer's eyes are directed a little to one
side of the illuminated surfaces, a match will no longer
exist; the mixed colour, which is to the audience
pinkish, will now appear a bluish green to him. The
reason for this alteration in hue is that the whole of the
images falls outside the yellow spot.

It will now be quite apparent that we must discount
any assertion in regard to colour matches, unless we are
told the distance of the eye from the surface on which

the match is made, together with the size of that surface. This yellow spot is often provokingly tiresome in the study of colour mixtures, and one might almost be justified in doubting whether any *absolutely* exact matches can ever be vouched for, owing to the important region of the retina which it occupies.

The fatigue of the retina to colour after it has been presented to the eye for any length of time is a difficulty, but in a less degree. That the retina does experience fatigue can be shown by a very simple experiment. The lecture theatre is now illuminated by the incandescent light, and if we throw an image of the bright carbon points of the electric arc light on the screen and steadily fix the eyes on the image of the white-hot crater for some (say) twenty seconds, and then we suddenly withdraw it, a *dark* image of the points will be seen on the partially lighted screen, and will appear to travel with the eyes as they move away from the fixed point. This phenomenon is due to the fact that the perceiving apparatus for white light gets fatigued on the parts of the retina on which the bright image of the white carbon points thrown on the screen fell, and that when the source of brightness was removed, the less intense illumination of the screen failed to stimulate the vision apparatus at those parts to the

same extent that they were stimulated over the rest of
the field. We can vary the experiment by placing a red
glass in front of the electric light, and, following the
same course as before, we shall see a greenish-blue
image of the carbon points upon the screen. In this
case the retinal apparatus which has not been stimulated by the red sensation will be capable of the
maximum stimulation by the feeble white light, whilst
that part which has suffered fatigue will not respond so
freely to the red contained in the white light. If we
abstract a certain amount of red from the spectrum,
its recombination will give a white tinged with greenish
blue, which is a counterpart of the colour we feel when
the eyes have been fatigued by the red light.

CHAPTER III.

LET me take you back again to matches of colour. We will now, however, make the matches with the primary colours in the guise of pigments. These colours themselves are complex colours, but as the eye cannot trace any difference, or at all events very little difference, between them and simple colours, a mixture of these complex colours should answer nearly as well as do mixtures of the simpler colours. We have here three discs, a red, a green, and a blue, and we can very closely match these colours by a red, a green, and a blue in the spectrum.

By having a radial slit cut to the centre of these card discs, we can slip one over the other so as to expose all three colours as sectors of a single disc. Then we can place the compounded disc on the axis of a rapidly rotating motor, and the colours will blend together, giving an uniform colour. Any proportions of the three colours can thus be mixed, and by a judicious alteration in them we now have them so arranged that they give a grey. By inter-locking together (Fig. 9) a black disc and a white disc, each with a diameter slightly larger than that of the other discs, but equal to each other, and rotating them on the same spindle behind the three colour discs, we can, by an alteration in the proportion of black to white, form a grey which will match that produced by the rotation of the three coloured sectors. In other words, white, though degraded in tone, can be produced by the three complex pigment

Fig. 9.

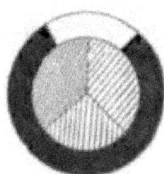

colours, as we have seen can also be done by the mixture of the three simple spectrum colours.

The mixture of the three spectrum colours can match other colours than white. For instance, it can be made to match the colour of brown paper. By the colour discs also we can do exactly the same by introducing, if necessary, a small quantity of white or black, or both, to dilute the colour or to darken its tone.

Another application of the same principles enables us to produce an artificial spectrum by means of a red, a green, and a blue glass. By fixing these three glasses behind properly shaped apertures cut in a card disc at

D

proper radial distances from the centre, and rotating the disc, we have upon the screen when light is passed through them a ring of rainbow colours. If the beam of light be first passed through a suitable rectangular aperture, the breadth of which is small compared with its length, placed close to the rotating disc, and an image of the aperture be focussed on the screen by a suitable lens, we shall have a very fair representation of the spectrum—every colour intermediate between the red and green, or the green and blue, being formed by mixtures of these pairs respectively.

We have now given a very fair proof that vision is really trichromic—that is, that it is unnecessary to have more than the sensations of three colours to produce the sensation of any of the others.

There is one colour, if it may be called so, that has not been shown you, and whether it is a simple colour or not cannot be stated. It seems, however, to be the basis of all other colours, since they all commence with it. It would, perhaps, be preferable to call it the first perception of light instead of a colour. We can exhibit this in a fairly easy manner by a little artifice. An incandescent lamp is before you, and a current from a battery passing through the carbon thread causes it to glow brightly. In the circuit, however, I have introduced what is known as a resistance, which consists of a very large number of square pieces of carbonized linen, pressed more or less tightly together. By means of a screw the pressure can be varied. When the pressure is somewhat relaxed, the resistance to the passage of the current is increased, and the carbon thread glows less brightly; and by a still greater release of pressure, the light can be made to disappear altogether. A beaker (Fig. 10) which we have here is covered with thin blotting paper, and when placed over the incandescent

FIG. 10.

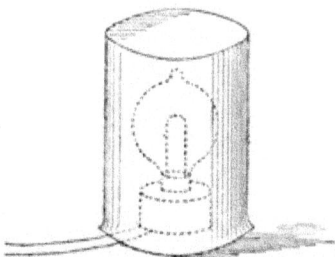

glow-lamp it appears as a luminous yellow cylinder, the colour being due to that of the light within it. We can next insert more resistance in the circuit, and it becomes red, due to the ruddy light of the thread. By inserting still more resistance into the circuit the red fades away, but in the darkness of this lecture theatre the beaker is still a luminous object, though faintly so. It has no colour, and the only sensation it provokes is one of light. Taking off the beaker, we see that the carbon thread is a dull *red* and nothing more. The passage of this light through the white blotting paper so reduces it that the red is non-existent, and the initial sensation is all we perceive.

Placing a piece of red, green, or blue gelatine round the lamp, we get the same effect, showing that the basis of all colour, be it red, green, or any other colour, is what appears to us to be colourless. This experiment is one which is full of interest, as it has a very distinct bearing on diagnosing our colour sensations, and a variation of it will have to be repeated under other conditions.

To go back, however, a little way, how does it arise that only three sensations are necessary to give the impression of all colours? One can understand that some definite period of the ether waves might be in unison with the possible swing of one apparatus in the eye, and another with another, but it is somewhat difficult at first sight to conceive that more than one can be made to answer to wave motion of a period with which it is out of tune, so to speak. A couple of illustrations taken from physical experiments may help to suggest how this can happen.

Fig. 11 is a double pendulum arranged as shown. The pendulum A is heavily weighted, whilst the pendulum B is light, being only a string with a small weight attached. This difference in weight was made designedly, to prevent any great effect of the movement of B being shown on A, though that of A must necessarily exercise a great influence on B. The two pendulums are now of the same length. A is set in motion, and as it swings, B also begins to swing, and soon is oscillating with greater motion than A, and continues to do so. The length of

Fig. 11.

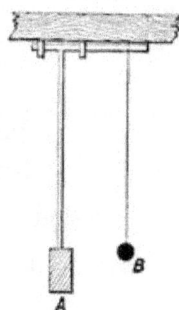

the pendulum B is next shortened, and A is again set in motion. B takes up the motion, and increases its swing more and more, but now the two pendulums are in opposite phases, and the motion of A tends to diminish the swing of B, and continues to do so till, after an interval of time, B is once more at rest, when it again will start swinging. The fact is, that when A commences to swing, B also commences; and as long as B and A are moving in the same direction the impulses tend to make B increase its swing, but when they are moving in the opposite direction, or rather, perhaps it should be said, when A begins to start from the highest point of its swing downwards whilst B is travelling upwards, the swing of B will gradually diminish. This, of course, must happen when B is shorter or longer than A, since their times of oscillation are then different. We can now picture to ourselves that when in the perceiving apparatus in the retina the moving parts—probably molecules or atoms— arrive at a certain amplitude, there is then an impression of light, and that it is quite possible that not only those waves whose motion is exactly of the same period as that of the apparatus will set them in motion, but also those waves which are actually of a very different period. If such be the case, it can be seen that waves of light of some periods may set each of the three kinds of perceiving apparatus in motion, and that possibly the resulting impressions given by the sum of all three for a wave out of tune with any of them may be even greater than when the wave period is absolutely the same as one of them. For in the last case a maximum effect may be produced on one apparatus, and the effects on the other two may be insignificant; whilst in the first case the effects on two of them may be so large that their combined effects may have a larger value.

The following diagram (Fig. 12), made on the principle of Lissajou's figures, shows graphically the motion of the pendulum. The pendulum, with a pen attached, was started by an independent pendulum, which had a

different period, and the amplitude of the former registered itself on paper which moved by clockwork round the axis of suspension. As the two pendulums had different periods, the amplitude, as shown by the traces made, first increased and then diminished till there was no motion, and then started again. The trace is very instructive, and deserves attention. It will be noticed that the amplitude, or length of swing, increased rapidly at first, and then very gradually attained a maximum.
Having attained this maximum, the amplitude diminished very slowly for some time,

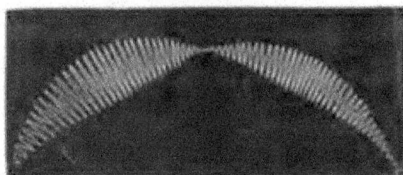

FIG. 12.

and finally came rather rapidly to zero, and the pendulum for an instant was at rest.

With the notion in our minds that the perceiving apparatus might act in the way that the pendulum acts, we naturally apply it to the theories which early investigators on colour vision propounded. Thomas Young, whose name has already been mentioned, had propounded a theory of vision, which depended on the existence of only three colour sensations, and Von Helmholtz adopted it and explained the action of the three sensations in reference to the spectrum as

shown in the diagram. These figures do not pretend
to be absolute measures of the sensations, but only
of the form which they might take (Fig. 13). The
height of the curve at each part of the spectrum is
supposed to represent the stimulation given to each
apparatus by the different colours. Looking at the
figures we see that each sensation has a place of

FIG. 13.

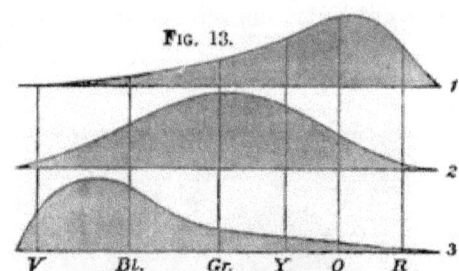

V   Bl.   Gr.   Y   O   R

The top figure is the red sensation on the Young
theory; the middle is the green sensation, and
the lowest the violet or blue sensation.

maximum stimula-
tion, and that the
stimulation falls
off more or less
rapidly on each
side of this maxi-
mum. It will, how-
ever, be noticed
that whilst the green sensation takes very much the
form of the pendulum amplitudes (Fig. 12) between its
periods of rest, the other two differ from it. In the case
of the red sensation, the stimulation falls very rapidly
in the red as it reaches the limit of visibility of the
spectrum, and in that of the blue sensation the steep
descent is towards the extreme violet. When the three
sensation theory is examined in the light of the careful
measurements that have been made, the results tell us
that these diagrams can only be taken as suggestive.

CHAPTER IV.

An independent investigator of this subject was Clerk Maxwell, who experimented with a "colour-box" of his own design, by which he mixed the simple colours of the spectrum, and the results he got are really the first which are founded on measurement. He measured something, but hardly arrived at the colour sensation. His colour-box took two forms, both on the same principles, so only one will be here described, the diagram and description being taken from his classic paper in the Philosophical Transactions of the Royal Society for 1860.

"The experimental method which I have used consists in forming a combination of three colours belonging to different portions of the spectrum, the quantity of each being so adjusted that the mixture shall be white, and equal in intensity to a given white. Fig. 14 represents the instrument for making the observations. It consists of two tubes, or long boxes of deal, of rectangular section, joined together at an angle of about 100°.

"The part A K is about five feet long, seven inches broad, and four deep; K N is about two feet long, five inches broad, and four deep; B D is a partition parallel to the side of the long box. The whole of the inside of the instrument is painted black, and the only openings are at the end A C, and at E. At the

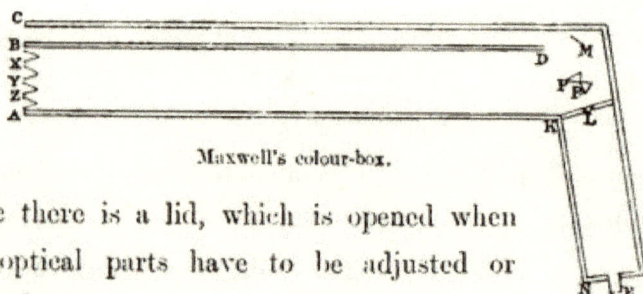

Maxwell's colour-box.

angle there is a lid, which is opened when the optical parts have to be adjusted or cleaned.

"At E is a fine vertical slit, L is a lens; at P there are two equilateral prisms. The slit E, the lens L, and the prisms P are so adjusted, that when light is admitted at E, a pure spectrum is formed at A B, the extremity of the long box. A mirror at M is also adjusted so as to reflect the light from E, along the narrow compartment of the box to B C.

"At A B is a rectangular frame of brass, having a rectangular aperture of six inches by one. On this frame are placed six brass sliders, X Y Z. Each of these carries a knife-edge of brass in the plane of the surface of the frame.

"These six movable knife-edges form three slits, X Y Z, which may be so adjusted as to coincide with any three portions of the pure spectrum formed by light from E. The intervals behind the sliders are closed by hinged shutters, which allow the sliders to move without letting light pass between them.

"The inner edge of the brass frame is graduated to twentieths of an inch, so that the position of any slit can be read off. The breadth of the slit is ascertained by means of a wedge-shaped piece of metal, six inches long, and tapering to a point from a width of half an inch. This is gently inserted into each slit, and the breadth is determined by the distance to which it enters, the divisions on the wedge corresponding to the 200th of an inch difference in breadth, so that the unit of breadth is ·005 inch.

"Now suppose light to enter at E, to pass through the lens, and to be refracted by the two prisms at P, a pure spectrum, showing Fraunhofer's lines, is formed at A B, but only that part is allowed to pass which falls on the three slits, X Y Z. The rest is stopped by the shutters. Suppose that the portion falling on

X belongs to the red part of the spectrum; then, of the white light entering at E, only the red will come through the slit X. If we were to admit red light at X, it would be refracted to E, by the principle in optics that the course of the ray may be reversed.

" If, instead of red light, we were to admit white light at X, still only red light would come to E; for all other light would be either more or less refracted, and would not reach the slit at E. Applying the eye at the slit E, we should see the prism P uniformly illuminated with red light, of the kind corresponding to the part of the spectrum which falls on the slit X, when white light is admitted at E.

" Let the slit Y correspond to another portion of the spectrum, say the green; then if white light is admitted at Y, the prism, as seen by an eye at E, will be uniformly illuminated with green light; and if white light be admitted at X and Y simultaneously, the colour seen at E will be a compound of red and green, the proportions depending on the breadth of the slits and the intensity of the light which enters them. The third slit Z, enables us to combine any three kinds of light in any given proportions, so that an eye at E shall see the face of the prism at P, uniformly illuminated with the colour resulting from the combination of the three. The position of these three rays in the spectrum is found by admitting the light at E, and comparing the position of the slits with the position of the principal fixed lines; and the breadth of the slits is determined by means of the wedges.

"At the same time, white light is admitted through B C to the mirror of black glass at M, whence it is reflected to E, past the edge of the prism at P, so that the eye at E sees through the lens a field consisting of two portions, separated by the edge of the prism; that on the left hand being compounded of three colours of the spectrum refracted by the prism, while that on the right hand is white light reflected from the mirror. By adjusting the slits properly, these two portions of the field may be made equal, both in colour and brightness, so that the edge of the prism becomes almost invisible.

"In making experiments, the instrument was placed on a table in a room moderately lighted, with the end A B turned towards a large board covered with white paper, and placed in the open air, so as to be uniformly illuminated by the sun. In this way the three slits and the mirror M were all illuminated with white light of the same intensity, and all were affected in the same ratio by any change of illumination ; so that if the two halves of the field were rendered equal when the sun was under a cloud, they were found nearly correct when the sun again appeared. No experiments, however, were considered good unless the sun remained uniformly bright during the whole series of experiments.

"After each set of experiments light was admitted at E, and the position of the fixed lines D and F of the spectrum was read off on the scale at A B. It was found that after the instrument had been in use some time these positions were invariable, showing that the eye-hole, the prisms, and the scale might be considered as rigidly connected."

With this instrument he made mixtures of three colours, to match with white. By shifting the slits into various positions and taking as his three standard colours a red near the C line, a green near E, and a blue between F and G (see frontispiece), he obtained a variety of matches, from which he formed equations. After eliminating, or rather reducing the errors to the most probable value by the method of least squares, he got from his matches with white a table of colour values in terms of the three standard colours, from which the diagram of the spectrum (Fig. 15) was made. (The

heights of the dotted curves are derived from the
widths of the slits, and the continuous curve is the
sum of these heights.) Now what appears to be a
properly chosen colour does not necessarily stimulate
only one sensation. Indeed the probabilities are against
it, except in the extreme red and extreme violet. If

FIG. 15.

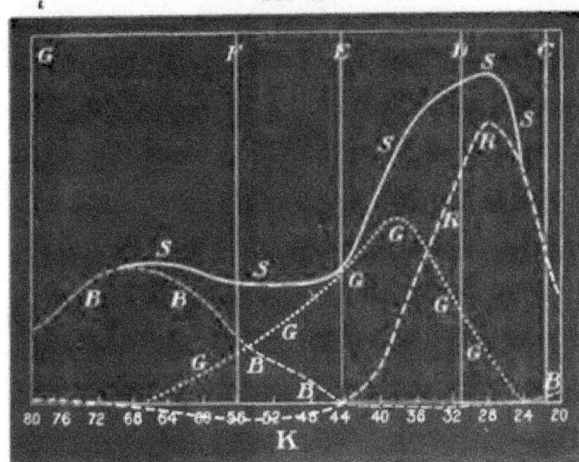

colours intermediate to the standard colours be
matched by a mixture of the latter, we do not arrive
at any solution of the amount of stimulation of each
sensation, since the chosen standard colours themselves
may be due to a stimulation of all three sensations.
As a matter of fact, Clerk Maxwell chose colours
which do not best represent the colour sensations.
The red is too near the yellow, as is also the green.

The blue should also be nearer the violet end of the spectrum than the position which he chose for it. We may take it, then, that except as a first approximation, Clerk Maxwell's diagrams need not be seriously taken into account. The diagram itself shows that the colour *sensations* are not represented by the colours he chose. Supposing any one in whom the sensation of green is absent were examining the spectrum, there would, according to the diagram, be no light visible at the green at E. Anticipating for a moment what we shall deal with in detail shortly, it may be stated that in cases where it is proved that a green sensation is absent, there is no position in any part of the spectrum where there is an absence of light. Had he chosen any other green, the same criticism would have been valid. The diagram as it stands is really a diagram of *colour mixtures* in terms of three arbitrarily chosen colours, and not of colour *sensations*. It merely indicates what proportions were needed of the three colours, which he took as standards, to match the intermediate spectrum colours. The negative sign in some of the equations—given in the appendix, page 201 —may be somewhat puzzling to those who have not made colour matches, but not to those who have actually made experiments. It means that where it is present no match of colour by a mixture of the standard colours is possible; and that it would be only possible if a certain quantity of the colour to which is attached a negative sign were to be abstracted—an impossible condition to fulfil, but one which may often occur in colour-matching experiments. Later you

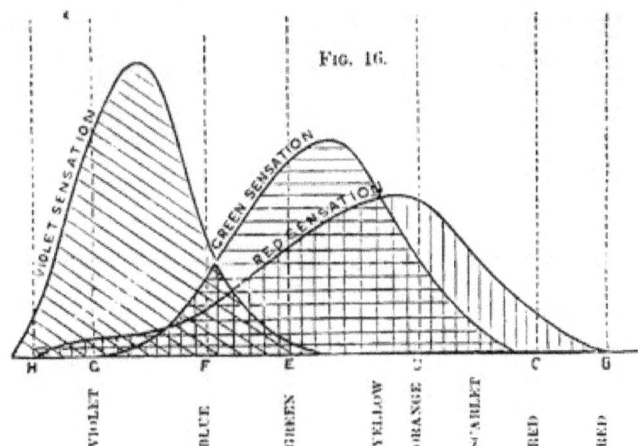

Fig. 16.

will find that when colours are chosen as standards so that the resulting equations give no negative sign for any colour, we have a criterion as to the colours which give the nearest approach to the true sensations. The next diagram (Fig. 16) of colour sensations is due to Kœnig, who investigated the subject with Von Helmholtz. By a modified method, which perhaps need not be explained in detail here, he produced them, and they must be apparently not far from the actual

E

state of things, supposing this theory be proved to be true. For my own part, I am under the impression that the positions of the colours which most nearly approach the colour sensations might be slightly altered in regard to the green and the blue, for reasons that will subsequently be given when the later experiments of General Festing and myself come to be described. For the immediate purpose of the lecture, the curves are sufficiently accurate, and I will ask you to notice what they tell us. It is presupposed in these diagrams that, if the three colour-perceiving apparatus are equally stimulated, a sensation of white will be produced; and the reverse, of course, is true, in that white will give rise to equal stimulation of the three apparatus. It follows, then, that in the parts of the spectrum where all three curves of sensation are seen to take a part in the production of a colour, such as at the E line, the colour is really due to the extra stimulation of one or two of the apparatus above that required to produce a certain amount of white. *The colour in every part of the spectrum may be represented by not more than two sensations, with a proportion of white.* In the orange and scarlet there are only two sensations excited, without any sensible amount of white, as the amount of violet sensation is extremely small. At the extreme ends of the spectrum only one sensation—the red or the violet—is excited; but in the region of the green the colour must be largely diluted with the sensation of white. As an example, we may take the part of the spectrum where the red and the violet sensation curves cut each other. At this point the green sensation curve rises higher than the intersection of the other curves. The red and the violet sensations have only to be mixed with an equal amount of the green sensation to make white, so that the height of the green sensation curve above the point of intersection represents the amount of pure green sensation which is stimulated. The colour is therefore caused by the green sensation, largely diluted with white. A scrutiny of the curves will show that at no point is the green sensation so free from any other as at this point, if we regard white by itself as a neutral colour. Looking at these figures, we can readily see what effect the removal of any one or two of the three sensations would have upon the colour vision of the individual. The probabilities, however, against two of the three sensations being absent must evidently be very much smaller than that there should be an absence of only one of the sensations, either red, green, or violet.

It will be well that we should also have before us the theory which is the only serious rival to that of Young, viz., that of Hering. In the report of the Colour Vision Committee there is an excellent description of this theory. As it was furnished by Dr. Michael Foster, we may be sure that the ideas of its originator are correctly given, and therefore I will quote it in his words :—

" Another theory, that of Hering, starts from the observation that when we examine our own sensations of light we find that certain of these seem to be quite distinct in nature from each other, so that each is something *sui generis*, whereas we easily recognise all other colour sensations as various mixtures of these. Thus, the sensation of red and the sensation of yellow are to us quite distinct ; we do not recognise anything common to the two, but orange is obviously a mixture of red and yellow. Green and blue are equally distinct from each other and from red and yellow, but in violet and purple we recognise a mixture of red and blue. White again is quite distinct from all the colours in the narrower sense of that word, and black, which we must accept as a sensation, as an affection of consciousness, even if we regard it as the absence of sensation from the field of vision, is again distinct from everything else.

Hence the sensations caused by different kinds of light or by the absence of light, which thus appear to us quite distinct, and which we may speak of as ‘native’ or ‘fundamental’ sensations, are white, black, red, yellow, green, blue.   Each of these seems to us to have nothing in common with any of the others, whereas in all other colours we can recognise a mixture of two or more of these.   This result of common experience suggests the idea that these fundamental sensations are the primary sensations, concerning which we are enquiring.   And Hering's theory attempts to reconcile, in some such way as follows, the various facts of colour vision with the supposition that we possess these six fundamental sensations.   The six sensations readily fall into three pairs, the members of each pair having analogous relations to each other.   In each pair the one colour is complementary to the other—white to black, red to green, and yellow to blue.   Now, in the chemical changes undergone by living subjects, we may recognise two main phases, an upward constructive phase, in which matter previously not living becomes living, and a downward destructive phase, in which living matter breaks down into dead or less living matter.   Adopting this view, we may, on the one hand, suppose that rays of light, differing in their wave-length, may affect the chemical changes of the visual substance in different ways, some promoting constructive changes (changes of assimilation), others promoting destructive changes (changes of dissimilation) ; and on the other hand, that the different changes in the visual substance may give rise to different sensations.

" We may, for instance, suppose that there exists in the retina a visual substance of such a kind that when rays of light of certain wave-lengths—the longer ones, for instance, of the red side of the spectrum—fall upon it, dissimilative changes are induced or encouraged, while assimilative changes are similarly promoted by the incidence of rays of other wave-lengths, the shorter ones of the blue side. But it must be remembered that in dealing with sensations it is difficult to determine what part of the apparatus causes them ; we may accordingly extend the above view to the whole visual apparatus, central as well as peripheral, and suppose that when rays of a certain wave-length fall upon the retina, they in some way or other, in some part or other of the visual apparatus, induce or promote dissimilative changes, and so give rise to sensations of a certain kind, while rays of another wavelength similarly induce or promote assimilative changes, and so give rise to a sensation of a different kind.

" The hypothesis of Hering applies this view to the six fundamental sensations spoken of above, and supposes that each of the three pairs is the outcome of a particular set of dissimilative and assimilative changes. It supposes the existence of what we may call a red-green visual substance of such a nature that so long as dissimilative and assimilative changes are in equilibrium, we experience no sensation; but when dissimilative changes are increased, we experience a sensation of (fundamental) red, and when assimilative changes are increased, we experience a sensation of (fundamental) green.

" A similar yellow-blue visual substance is supposed to furnish, through dissimilative changes a yellow, through assimilative changes a blue sensation; and a white-black visual substance similarly provides for a dissimilative sensation of white and an assimilative sensation of black. The two members of each pair are therefore not only complementary but also antagonistic. Further, these substances are supposed to be of such a kind that while the white-black substance is influenced in the same way, though in different degrees, by rays along the whole range of the spectrum, the two other substances are differently influenced by rays of different wave-length. Thus, in the part of the spectrum which we call red, rays promote great dissimilative changes of the red-green substance with comparatively slight effect on the yellow-blue substance; hence our sensation of red.

"In that part of the spectrum which we call yellow, the rays effect great dissimilative changes of the yellow-

Fig. 17.

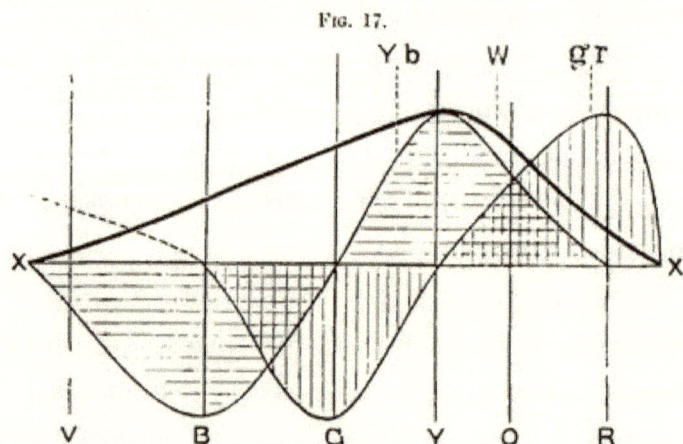

blue substance; but their action on the red-green substance does not lead to an excess of either dissimilation or assimilation, this substance being neutral to them; hence our sensation of yellow. The green rays, again, promote assimilation of the red-green substance, leaving the assimilation of the yellow-blue substance equal to its dissimilation; and similarly blue rays cause assimilation of the yellow-blue substance, and leave the red-green substance neutral. Finally, at the extreme blue end of the spectrum, the rays once more provoke dissimilation of the red-green substance, and by adding red to blue give violet. When orange rays fall on the retina, there is an excess of dissimilation of both the red-green and the yellow-blue substance; when greenish-blue rays are perceived, there is an excess of assimilation of both these substances; and other intermediate hues correspond to various degrees of dissimilation or assimilation of the several visual substances. When all the rays together fall upon the retina, the red-green and yellow-blue substances remain in equilibrium, but the white-black substance undergoes great changes of dissimilation, and we say the light is white."

It has been said by the same writer that this theory is tri-chromic. For my own part I do not see that it is so in the sense in which that word is used in the theory of Young. It may be a tetra-chromic, for as far as *colour* is concerned the black-white sensation must be excluded; but it appears to me that it cannot be strictly brought under the head of tri-chromic.

CHAPTER V.

THE readiest means of investigating the stimulation of the different sensations necessary to produce colour is evidently by eyes in which one or two sensations are absent, and this applies not only to the Young theory, but also to that of Hering.

In former days, not much more than a century ago, the existence of colour blindness, as it is now named, was a matter of great curiosity, and in the Philosophical Transactions of the Royal Society of 1777, the case of a shoemaker named Harris is described by a Mr. Huddart, who travelled all the way from London to the Midlands on purpose to see if all the alleged facts regarding the patient were true. Harris mistook orange for green, brown he called black, and he was unable to distinguish between red fruits and the surrounding green leaves. At first, colour blindness was called Daltonism, from the fact that the great chemist Dalton suffered from it, and investi-

gated the variation which he found existed in his
vision from that of the majority of his fellow-creatures.
It was in 1794 that Dalton described his own case of
colour blindness. He was quite unaware of his defect
till 1792, when he was convinced of its existence
from his observations of a pink geranium by candle-
light. "The flower," he says, "was pink; but it
appeared to me almost an exact sky-blue by day. In
candle-light, however, it was astonishingly changed,
not having any blue in it; but being what I call a
red colour which forms a striking contrast to blue."
He goes on to remark that all his friends except his
brother (mark this relationship), said: there was not
any striking difference in the two colours by the two
lights. He then investigated his case by the solar
spectrum, and became convinced that instead of having
the normal sensations, he only had two or at most
three. These were yellow, blue, and perhaps purple.
In yellow, he included the red, orange, yellow, and
green of others, and his blue and purple coincided
with theirs. He says, that "part of the image which
others call red, appears to me little more than a shade
or defect of light; after that, the orange, yellow and
green seem *one* colour, which descends pretty uniformly
from an intense and a rare yellow, making what I

should call different shades of yellow. The difference
between the green part and the blue part is very
striking to my eye, they seem to be strongly con-
trasted. That between the blue and purple much less
so. The purple appears to be blue much darkened
and condensed."

Dalton said a florid complexion looked blackish-blue
on a white ground. Blood looked like bottle green,
grass appeared very little different from red. A laurel
leaf was a good match to a stick of sealing-wax.
Colours appeared to him much the same by moonlight
as they did by candle-light. By the electric light
and lightning, they appeared as in day light. Some
browns he called red, and others black.

Mr. Babbage, in Scientific London (1874), gives an
account of Dalton's presentation at Court.

Firstly, he was a Quaker, and would not wear a
sword, which is an indispensable appendage to
ordinary Court-dress. Secondly, the robe of a Doctor
of Civil Laws was known to be objectionable on account of its colour—scarlet, being one forbidden by
the Quakers. Luckily, it was recollected that Dalton
was affected with that peculiar colour blindness which
bore his name, and that as cherries and the leaves of a
cherry-tree were to him of the same colour, the scarlet

gown would present no extraordinary appearance. So
perfect evidence was the colour blindness, that the
most modest and simple of men, after having received
the Doctor's gown at Oxford, actually wore it for
several days in happy unconsciousness of the effect
he produced in the street. The rest of the description
we need not reproduce. Both the above cases we shall
see shortly come under the category of red-blindness
in the Young theory. Recent investigations tell us
that such colour blindness is by no means rare, nor can
it have been then. Statistics, derived from carefully
carried out examinations made in various parts of the
world by an approved method of testing, show that
about four out of every hundred males suffer from
some deficiency in colour perception, but that so far
as the more limited statistics regarding them are to
be depended upon, only about four out of every 1000
women suffer in the same manner.

Colour blindness in a healthy subject is usually
hereditary, and is always congenital. It is curious
to trace back in some instances the colour blindness,
where it is to be found, in a family. It often happens
that colour blindness—as the gout is said to do—
skips a generation. This is usually traced to the
fact that the generation skipped is through the

mother. Thus, the maternal grandfather may be colour blind, as may be the grandsons, but the mother will very frequently have perfectly normal vision for colour. On the other hand, the paternal grandfather may have defective colour perception, and this may be inherited both by the grandsons and the father. The remark made by Dalton regarding his brother's eyesight points to the fact that his own colour blindness was probably hereditary. Deaf mutes, Jews and Quakers, seem to be more liable to colour blindness than other people, statistics giving them 13·7, 4·9, and 5·9 as the percentages. It may be well to point out that the deficiency in colour perception to which we are alluding is totally distinct from that which may arise from disease. This last form has such marked characteristics of its own that it can at once be distinguished from the congenital form.

Of the four per cent. of males who suffer from congenital colour deficiency of vision, a large number are not totally lacking in any one or more colour sensations. Those in which one sensation, on the Young theory, is entirely missing are called "completely red-, green-, or violet-blind," whilst those in which the sensation is but partially deadened are called "partially red-, green-, or violet-blind." When

two sensations are entirely absent, and such cases
are very rare indeed, they are generally said to have
monochromatic vision ; that is, every colour to them
is the same, as is also white, the only distinction
between any of them being the superior brightness of
some over others. The best illustration of this form
of colour vision is perhaps to say that the retinæ of
such people have the same characteristics in regard
to sensitiveness as has a photographic plate, the
resulting prints in black and white representing what
it sees in nature. When we have to adopt the
terms used by the followers of Hering's theory—
the theory which obtains most followers amongst
the physiologists, since it endeavours to explain
colour vision in a physiological way, though it fails
to meet all the requirements of the physicist—we
should restrict our terms to red-green and yellow-
blue blindness, still perhaps retaining the term monochromatic vision for the rare cases specified above.
As we must employ some terms to express our mean-
ing, we shall in these lectures adopt those of the
Young theory.

Now taking a red-blind person and examining him
with the spectrum, we find that he sees no light at
all at the extreme limit of our red, and only when

he comes to the part where the red lithium line marks
a certain red does a glimmer commence ; he then sees
what he may call dark-green, or he may call dark-
yellow. When questioned about what to us are greens
he also calls them green or yellow, some being bright,
others saturated hues, and others again paler. When
he gets to the bluish-green he calls it grey, and will
say it is indistinguishable from, and in fact will match
with, a white degraded in tone. From this point he
will say he sees blue, near F pale-blue, and in the violet
dark-blue. Too much importance must not be attached
to the nomenclature adopted by the colour blind. They
have to take the names of the colours from the normal
eyed. Yellow objects are generally brighter than red,
and having annexed the idea that what to them is
bright red is called yellow, they give it that distinguishing name. His limit of vision at the violet
end will be the same as the majority of mankind,
but it will be considerably shortened at the red end.
The point in the spectrum which he calls grey is an
important point, and corresponds to the place where
the violet and green curves cut in Fig. 16. This
point can be very accurately determined by placing
alongside the colour patch A (Fig. 6) the white
patch, which is reduced in brightness as required by

rotating sectors. As the slit is moved along the spectrum it will eventually reach a point where he will say both patches of light are exactly similar in hue. To the normal eye one will be white and the other the kind of green indicated above (see frontispiece).

If a similar examination be made of the green-blind, the red end of the spectrum will be called red or yellow, but the spectrum itself will be visible between the same limits as it is to the person who has the normal sense of vision. A grey stripe will be seen in the spectrum, but in this case it will be a trifle nearer the red end of the spectrum than the point which the red-blind calls grey; from this point to the extreme violet, the green-blind will name the spectrum colours similarly to the red-blind. The part of the spectrum where grey exists to the green-blind is even more important than that part at which it exists to red-blind, for it marks the place where the red and violet curves cut each other in Fig. 16, and is in the majority of cases the place in the spectrum where to the normal eye the green sensation is unmixed with any sensation except that of white, as quite recently explained. This green evidently is the colour which is most usefully employed in making colour mixtures in order to obtain the three

F

sensation curves of the Young theory, since white can be added to the colour matched. To avoid verbiage, we shall call the points where the red- or green-blind see a grey in the spectrum their neutral points, and the grey they see at those points their neutral colours. The three curves we shall call the red, green, or violet curves, and the slits, when placed in the red, green, or violet of the spectrum, as the red, green, and violet slits.

We have already mentioned the case of those who possess monochromatic vision, and shown in what respect they will differ in their description of the spectrum from those more common cases of defective vision. If the visual sensation they possess be the violet, they will see no light at the extreme red of the spectrum, and very little in the orange. They must match every colour with some shade of grey, for they will only perceive that sensation, in what to ordinary normal eyes is white. We need not detail how those who possess monochromatic vision due to some other sensation would describe the different colours. The diagram will tell us. Suffice it to say, that one colour will only differ from another and from white in brightness.

It is a very remarkable fact how many people who

are defective in colour vision pass through a good part of their lives without being definitely aware of it. It is very doubtful whether, in the majority of cases, they themselves discover it. They may quite possibly attribute the descriptions of colour which they hear, and which appear to them absolutely false or meaningless, as due to mental or moral defects in their friends. I have had two cases of this recently. One was a gentleman of seventy-four, who had no conception that he had anything but normal colour vision; his daughters, however, had a suspicion that something was not quite right in it, and after a good deal of persuasion brought him to me to examine. The first mistake that he made was to state that he was sitting on a black velvet chair, whereas the seat was a deep crimson plush. He laughed at his daughter's description of the mistake he made, and declared he was only colour ignorant, and that she was the one who was colour blind. The examination showed that colour ignorant he was, but that the ignorance was due to complete red-blindness. For the seventy-four years he had lived he was unaware of his deficiency, suspecting it in others, and it was only an accidental circumstance which made him acquainted with the true state of his colour percep-

tion. Another elderly gentleman, in a high position
in life, was also accidentally tested, and he proved
to be completely green-blind. He, too, was quite
unaware of his defect, and protested that, yachtsman
as he was, he would never mistake a ship's lights;
but a very brief test showed his friends who were
with him that his declaration had to be received
with a certain amount of reservation. Others there
are who certainly do know that some peculiarity
exists in their sense of colour, and, foolish as it may
appear to be—though, after all, it is quite consistent
with a sensitive nature—they have tried to hide their
defect from their fellow-creatures. Such examples, no
doubt, some of my audience have met with, and
experience tells me that they have just as much
reluctance to pass an hour in my darkened room as
they would have to occupy a police cell. In those
few cases that have come voluntarily to me for
examination, the peculiarity in colour sense was first
brought to notice by the patient—if patient I may
call him—failing to distinguish between cherries and
the cherry leaves, or strawberries and the strawberry
leaves. Such mistakes committed publicly are usually
the source of unbounded merriment and curiosity to
schoolboys when made by their schoolfellows, and I

am bound to say that even persons of graver years are
not unapt to be amused at what they consider to be
a shortcoming in their fellow-creatures. To the student
of colour vision the discovery of curious cases of colour
deficiency is looked upon in a very different light
—a good case of colour blindness, or still better one of
monochromatic vision, is eagerly sought after, with the
hope of submitting it to a rigid examination. When

FIG. 18.

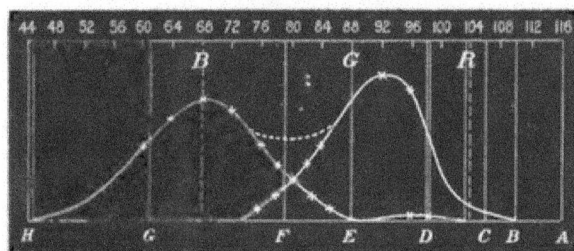

we look at the diagram (Fig. 16) we shall find why it
is that the colour blind describe the spectrum as they
do. Literally for those whose vision is di-chromic, it is
made up of two sensations alone, and the colours to
which these sensations give rise are mixed throughout
a large part of the spectrum, the pure unmixed sensations being at each end of the spectrum as they are in
normal colour vision. The annexed diagram (Fig. 18)
gives the curves for a red-blind person as made by
observations under Clerk Maxwell's directions. The

standard colours here have been badly selected, for one of them stimulates the two sensations possessed.

An easy and instructive experiment can be made to give an idea of the kind of colour that these colour blind imagine as white, whether they be red-, green-, or violet-blind. (For those who have only monochromatic vision, as before stated, white is coloured with the one colour they possess.) Three slits are now in the spectrum, one near the extreme end of the red, another well in the violet, and the third in that part of the spectrum in which the green-blind see their neutral colour (see page 66). With the three colours issuing from these apertures a match is made with the white patch, and in this case the match is made as seen from a distant point, so that the resulting deductions may be true to the audience. If a colour-blind person be in this theatre, he will agree with me that the match is as correct to him as it is to myself and the rest of you. So far we could not distinguish his colour perception from the normal, but if he be red-blind, and the red slit be covered, he will still say that the match holds good, for, as a matter of fact, the red with which we helped to build up the white is non-existent to him. The white that he now sees is to us the greenish-blue patch which the mixed violet and green make. If he be a green-blind person he will tell us the colour is a very pale blue, but when the green slit is covered up and the red uncovered, the match will once more be correct, though the purple, formed by the mixture of red and blue, will appear to him to be a little darker than the white. This is what one would expect, for you must recollect this green in the spectrum he would call white or grey. If then, from what to him is also white, though formed by the rays coming through the three slits, we take away a certain amount of degraded white (green to us), he must still see white, but darker. We have, however, met with what is an apparent paradox. The green, coming through the now covered slit, he calls white, as he also does the purple. To impress this point more strongly upon you, I will place in front of the green slit a small prism which has an angle of about one and a-half degrees. This is just sufficient to throw the green colour on the neighbouring white surface. Here we have both the colours which the green-blind calls white side by side. If the brightness of each be the same, he would see no difference in them. Is it possible that on any theory this can be correct? To explain this apparent paradox, and without reference to the mathematical proof that white subtracted from white leaves white, we have only to look at our diagram (Fig. 16), and it is immediately apparent how it arises. The red and the blue curves cut at this point; and if we take away the green sensation entirely, the residue will be a mixture of the red and blue, which is the identical purple colour forming the patch.

If we are wishful to ascertain the colour that the violet-blind calls white, we have only to cover up the violet slit and a yellow is left behind as the result. I would have you remark that these colours which are seen as white would only be of the hues shown you, supposing the colour sensations were identical with those in normal vision. Whether this is the case we cannot absolutely say, and the only way in which this can be authoritatively settled is by examining some person who has normal colour vision in one eye and defective colour sense, *not due to disease,* in the other. One such person has been examined abroad, but in what way I am unable to say. It is recorded that he sees the red end of the spectrum as yellow with the eye that is defective. Another person I have heard of in England, but so far have not had the good fortune to get hold of him for examination. When I can lay my hands on him, he will be able to help to confirm or disprove what should be a general rather than a particular case.

So far I have only met with what appears to be one genuine case of violet blindness. It is very remarkable, on account of the eccentricity of the colour nomenclature. The only two colours which the subject saw were red and *black*. He named all greens and blues as black, the distinction between the two being that the former was "bright black" and the latter "dark black." Yellow he called white, and a glance at Fig. 16 will show that at this place in the spectrum the neutral point of a violet-blind should occur. By shifting the slit gradually into the green, he called it grey, instead of "bright black," though it did not match the white patch when darkened. He called a green light a "bright black" light. We shall have to refer to this case when we are describing other investigations.

CHAPTER VI.

ANOTHER mode of exhibiting colour blindness, and one of the first adopted, is by making mixtures of colours with rapidly rotating colour discs. In my own experiments I have chosen a red, which is scarlet, over which a wash of carmine has been brushed. It has a dominant wave-length of 6300. The green is an emerald-green, and has a dominant wave-length of 5150. The blue is French ultra-marine, with a dominant wave-length of 4700. The card discs, of some 4 inches diameter, are coated with these colours as pastes, and by making an incision in them radially to the centre, as before described, and inter-locking them, the compound disc can be caused to show sectors of any angle that may be required. Outside these are the discs of black and white, the proportions of which can be altered at will.

The light thrown on the rotating sectors being

that from an electric arc light, normal vision requires 118° of red, 146° of green, and 96° of blue to match a grey made up of 75 parts of white and 285 parts of black. For the last two numbers a correction has been made to allow for the small amount of white light reflected from the black surface. This correction has also been made in the subsequent matches which will be described. Colour mixtures such as these are conveniently put in the form of equations, and that given will then be shown as follows—

118 R + 146 G + 96 U = 75 W + 285 B.

(Here R, G, U, W, and B are used to indicate Red, Green, Blue, White, and Black.)

This match was exact also for all the colour blind, for the deficiency in one grey is also a deficiency in the other. With a red-blind, however, very different matches can be made, as the red pigment is a complex colour. There is in it, besides red, a certain amount of yellow, whilst in the green there is, besides green, a small amount of a red and a larger amount of yellow. The yellow will not only stimulate the green sensation, but also the red where it is present. Although in complete red-blindness the red sensation is totally absent, we may expect that a mixture of red and blue, as well as of green and blue, will enable a match to be made of the grey produced by the mixture of white and black.

This was the case. We have the following proportions—   295 R + 65 U = 45 W + 315 B.

When the green disc is substituted for the red, the red-blind made the following mixture—

229 G + 131 U = 120 W + 240 B.

It is worth noticing that the amount of blue in the first mixture is about half that required for the second. This tells us that the amount of green sensation stimulated in the first case is much less than in the second. As red can be substituted for green, it should follow that green, when rendered darker, should match the red. To try this a red disc replaced the black disc, and a black disc replaced the blue. The following match was then made—

131 G + 229 B = 340 R + 20 W.

It seems impossible to believe that these mixtures, so dissimilar in colour, could ever form a satisfactory match. This last equation might have been derived from the two first, in which case it would have stood—

137 G + 223 B = 342 R + 18 W.

By a completely green-blind the following mixtures were made—

$$251 \text{ R} + 109 \text{ U} = 62 \text{ W} + 298 \text{ B,}$$
and
$$277 \text{ G} + 83 \text{ U} = 107 \text{ W} + 253 \text{ B.}$$

In this case 363 Green are equivalent to 251 parts of
Red mixed with 78 of White and 34 Black. The
difference in the matches made by the two types
of colour blindness is very evident. In the one
case the amount of red required is much greater
than the green, and in the other *vice versâ*. Another
instance may be given of colour matches made, by
means of discs, by a *partially* green-blind person,
whose case will be more fully described when we
treat of the luminosity of the spectrum to the different
classes of colour vision.

His matches were as follows — 1st, That of the
normal vision. 2nd,—

$$160 \text{ R} + 80 \text{ G} + 120 \text{ U} = 72 \text{ W} + 288 \text{ B.}$$

The green was then altered to 200, when the following
made a match—

$$65 \text{ R} + 200 \text{ G} + 95 \text{ U} = 72 \text{ W} + 288 \text{ B.}$$

Using these two equations, we have the following
curious result—that 120 G was matched by 95 R +
25 U. As the green disc is nearly twice as luminous
as the red to normal colour vision, this equation
confirms the result otherwise obtained, that his

blindness to colour is a deficiency in the green sensation. No mixtures of blue and red, or blue and green, would match a grey formed by the rotation of the black and white sectors.

I must now introduce to your notice a different method of experimenting with colour vision. If we throw the whole spectrum on the screen, and ask a person with normal vision to point out the brightest part, he will indicate the yellow, whilst a red-blind will say the green, and so on. This tells us that the various types of colour blind must see their spectrum colours with luminosity differing from that of the normal eye. The difference can be measured by causing both to express their sense of the brightness of the different parts of the spectrum in terms of white light, or of one another. Brightness and luminosity are here used synonymously. On the two small screens are a red and a green patch of monochromatic light—a look at the green shows that it is much brighter than the red. Rotating sectors, the apertures of which can be opened or closed at pleasure during rotation, are now placed in the path of the green ray. The apertures are made fairly small, and the green is now evidently dimmer than the red. When they are well open the green is once more brighter. Evidently at some time during the closing of the apertures there is one position in which the red and green must be of the same brightness, since the green passes through the stage of being too light to that of being too dark. By gradually diminishing the range of the "too open" to "too close" apertures we arrive at the aperture where the two colours appear equally bright. The two patches will cease to wink at the operator, if we may use such an unscientific expression, when equality in brightness is established. This operation of equalising luminosities must be carried out quickly and without concentrated thought, for if an observer stops to *think*, a fancied equality of brightness may exist, which other properly carried out observations will show to be inexact. Now, instead of using two colours, we can throw on a white surface a white patch from the reflected beam, and a patch of the colour coming through the slit alongside and touching it. The white is evidently the brighter, and so the sectors are placed in this beam. The luminosity of (say) a red ray is first measured, and the white is found to require a certain sector aperture to secure a balance in brightness. We then place another spectrum colour in the place of the first, and measure off in degrees the brightness of this colour in terms of white light, and we proceed similarly for the others. Now how are we to prove that the measures for luminosity of the different colours are correct? Let us place three slits in the spectrum, and by altering the aperture of the slits make a mixture of the three rays so as to form white. The intensity of this white we can match with the white of the reflected beam. We can then measure the brightness (luminosity) of the three colours separately, and if our measures are correct there is *primâ facie* reason to suppose that they will together make up the brightness of the white. Without going through this experiment it may at once be stated that the reasoning is correct, for within the limits of error of observation they do so. Having established this proposition, we can next compare *inter se*, the brightness of any or all of the rays of the spectrum by a preliminary comparison with the reflected beam of white light. As in the colour patch apparatus all colours and principal dark lines of the solar spectrum are known by reference to a scale, in making a graphic representation of the results, we first of all plot on paper a scale of equal parts, and at the scale number where a reading is made, the aperture of the sectors in degrees is set up. Thus, suppose with red light the scale number which marked the position of the slit was 59, and the aperture 10°, we should set up at that scale number on the paper a height of 10 on any empyric scale. If in the green at scale No. 38 the sectors had to be closed to 7°, we should set up 7 at that number on the scale.

When observations have been made at numerous places in the spectrum, the tops of these ordinates, as they are called, should be joined, and we then get the observed curve of luminosity for the whole spectrum. For convenience' sake we make the highest point 100, and reduce the other ordinates in proportion. For some purposes it may be advantageous to give the luminosity curve in terms of a scale of wavelengths. For our purpose, however, it is in general sufficient to use the scale of the instrument.

Now, if we test the vision of the various types of colour blind by this plan, we should expect to get luminosities at different parts of the spectrum which would give very different forms to these curves. We cannot hope, for instance, that a red-blind who sees no red in the extreme end of the spectrum would show any luminosity in that region, nor that the green-blind should show as much in the green part of the spectrum as those who possess normal colour vision, since one of the sensations is absent. With monochromatic vision there should be a still further departure from the normal curve. That these differences do exist is fully shown in Fig. 19. One of the most striking experi-

FIG. 19.

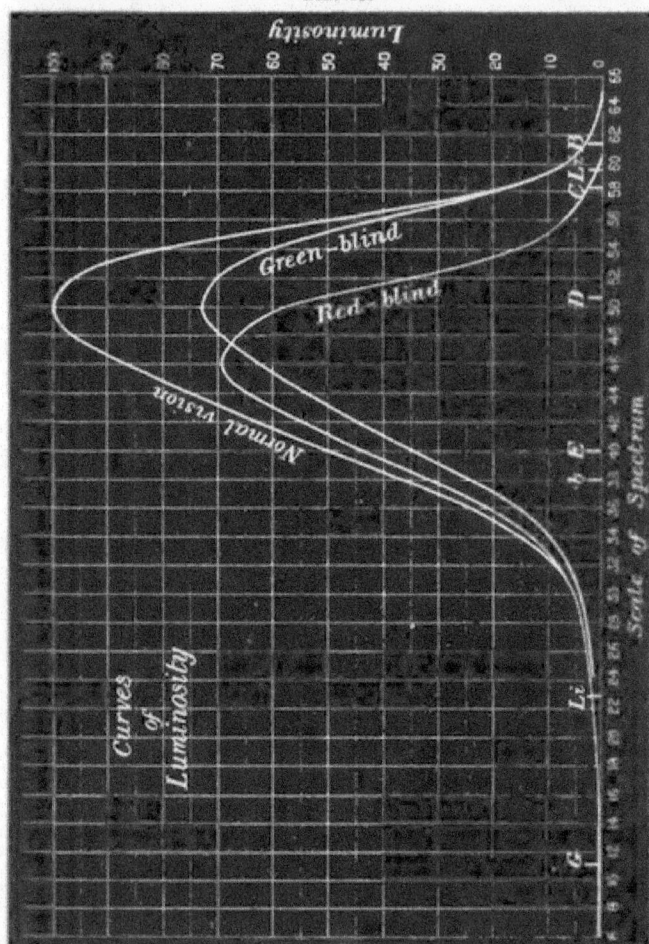

ments in colour vision is to place a bright red patch on the screen, and to ask a red-blind to make a match in luminosity with the white. The latter will have to be reduced to almost darkness—a darkness, indeed, that makes the match almost seem incredible. You will notice that the places in the spectrum where the red- and green-blind see grey are by no means places of greatest luminosity. We shall find that these luminosity curves are suggestive when making another investigation into the form of the spectrum curves of the colour sensations.

Besides cases of complete blindness due to the absence of one or two sensations on the Young theory, we have other cases, as was said when remarking on the percentage of people who are colour deficient, in which one or even two sensations are only more or less deadened. It has often been said that with the theory provisionally adopted, such cases are difficult to class as red or green deficient. As far as my own observations go, I have never found this difficulty. The luminosity curves of such observers, combined with other indications, give a ready means of classing them. The main difficulty to my mind is to state what is normal colour vision, but, as I have found that the very large majority of eyes give the same luminosity to colours as my own, I have taken my own colour perception as normal. In numerous experiments which Lord

Rayleigh has made in matching orange by means of a mixture of red and green, he has come across several who have apparently normal vision, as they see colours correctly in every part of the spectrum, and yet some require much less red mixed with the green to make a match with the orange than do others. What is yellow to them is decidedly green to the majority. This has been classed as another kind of normal vision ; but the luminosity curves show that it may be equally well due to a deficiency in the green sensation, and which would require more green to make the necessary match. The limits of the visible spectrum to these persons, as far as my examination of their cases goes, are the same as my own.

Again, there are others in which the spectrum seems decidedly somewhat shortened at the red end compared with my own, and the luminosity curves point to them as being strictly colour deficient in the red and nothing else. As they see all colours, they have been classed as another form of normal vision. The deficiency in both these cases is so small that white is their neutral colour, but there is evidence that the hues are slightly changed. I do not wish any one to accept my deductions as being more correct than those who hold differently, but the

results of examination by the luminosity methods
appear to me difficult to reconcile with any other view.
There are, however, a large number of cases in which,
though complete red- or green-blindness is wanting,
there is no doubt that more than slight colour
deficiency exists. For instance, in Fig. 23 we have
the curve of luminosity of the spectrum as measured
by a very acute scientific observer, and it is compared
with that of normal colour vision. He certainly is
not completely blind to any sensation. An inspec-
tion and comparison of the two curves will show
that he is defective in the green sensation, although it
is present to a large extent. The deficiency is obvious
enough. An endeavour to find his neutral point was
most interesting. At 39 in the scale he saw a little
colour, but at 39·5 all colour had vanished, and
between the coloured patch and the white he saw no
difference. This similarity he saw till 47·3 in the
scale, when he began to see a faint trace of colour.
There is a large piece of the spectrum, then, which
to him is grey. It must be recollected that all three
sensations were excited in this region, but some more
than others. Now, experiment has shown that, with
normal vision, two per cent. of any colour may be
mixed with a pure colour without its being perceived.

It is not surprising, therefore, that although the red, or the green, or the blue may be present in an in-

FIG. 23.

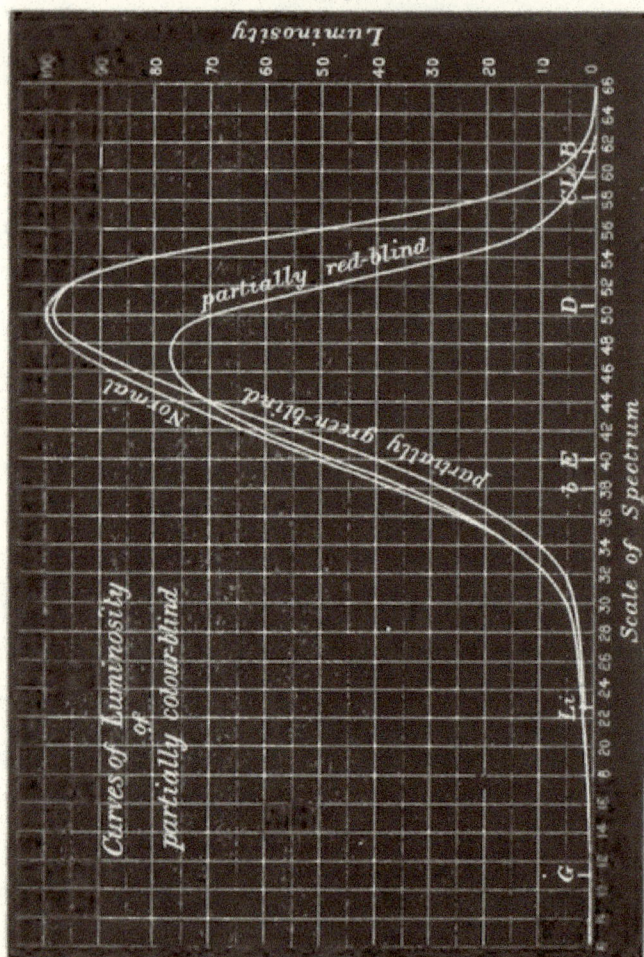

tensity above that required to form white, yet the resulting sensation should pass for white. It may be remarked that red and white when mixed he never mistook for yellow, and he always recognised yellows and red ; yellowish green, however, he called pale yellow.

Another example of partial red-blindness is also instructive. Fig. 23 also shows it graphically. There is no doubt as to the nature of the defect. The spectrum is slightly shortened, and the luminosity of this part of the spectrum is less than that of normal vision. There was no difficulty in distinguishing every colour, though the positions of the colours from yellow to green seemed to be shifted ; but no neutral point could be traced. Apparently, both this case and the former are about equally colour defective ; but in this last the same reasons do not apply for the existence of a neutral point. (For measures see page 214.)

CHAPTER VII.

WE are now in a position to carry the investigations
as to luminosity a little further. When we look at
small patches of light, we view the colour through the
yellow spot in the eye. If, when we have matched
the luminosity in the ordinary manner, we turn our
eyes some 10° away from the patches, we shall find
that except at one place in the green the equality in
brightness no longer exists. By a little practice we can
make matches of luminosity when the eyes are thus
diverted. This will give us a different curve of
luminosity, as the yellow spot absorption is absent,
and the difference in the heights of the ordinates
between the two curves will give us that absorption.
Fig. 20 shows this very well; and it will be noticed
that the eye is appreciably not so sensitive to the red
and yellow at 10° from the axis as it is on its central
area. If we measure the areas of these two curves we
get the relative values of the light energy which is

active on the two parts of the eye, and these we found
to be as 167 to 156. The heights at which to put the

FIG. 20.

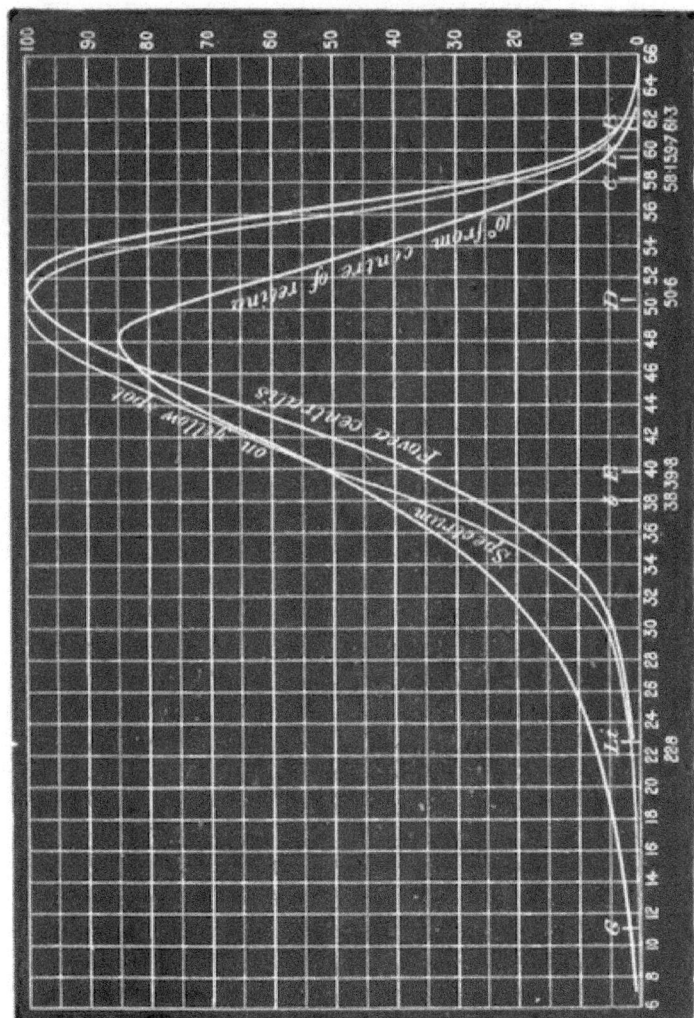

maxima of the two curves were found from various considerations, and the correctness of the deductions was verified by directly comparing the intensities of two patches of white light some 10° apart, which, when looked at direct, were of equal intensities. When one was compared with the other, the eye receiving one image centrally and the other outside the yellow spot, the difference in values was closely proportional to those of the above areas. The part of this last curve showing a deficiency in red sensation is very similar to that obtained from a person who is partially colour blind. The absorption by the yellow spot derived from these measures is graphically shown in the next figure (Fig. 21).

The question of the visual sensation at the "fovea centralis" (if it be admitted that this is coincident with the visual axis of the eye, as is usually accepted) may be very easily studied. When the luminosity of the spectrum is examined at five or six feet distance, by throwing the two patches on the whitened face of a small square of half-inch side, we get a result differing from both of the above. The fovea appears to be slightly more sensitive to red than the macula lutea, and is generally less sensitive to the green rays (see Fig. 20). If a star, or a distant light, be observed with

the part of the retina, on which the axis of the eye falls,
as is the case in ordinary vision, and then be observed

Fig. 21.

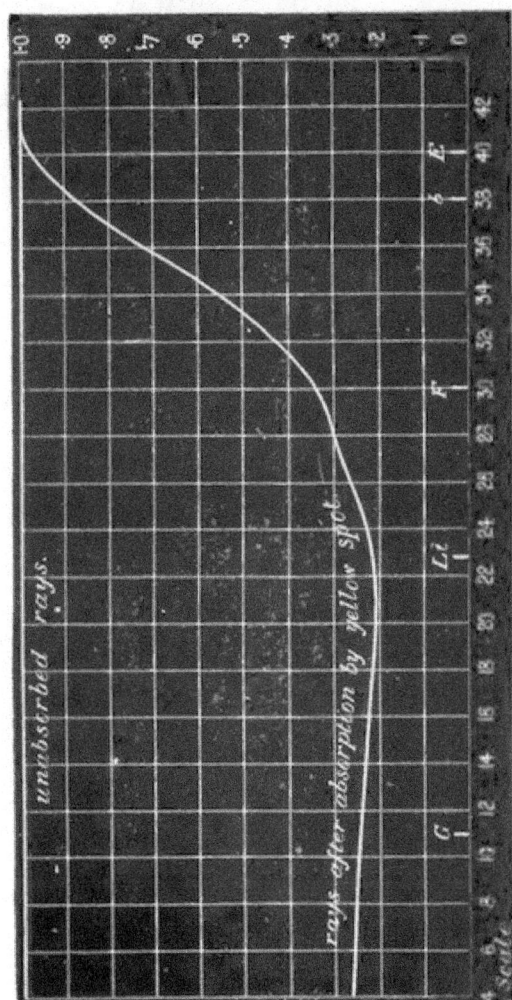

with the eye slightly directed away, the difference in the colours of the light is unmistakable. (The tables giving the measured value of these curves will be found in the appendix, page 211.)

Can we in any way find from these methods the colour sensation curves? I think we can. Suppose we have a second instrument exactly like the first placed side by side with it, we can then throw two patches of colour on the two adjacent white surfaces, and we can mix with either, or both of them, as much white light as we choose. From the second instrument let us throw all the spectrum colours in succession on to the one surface, and on to the other the three primary colours mixed in such proportions as to match them accurately. This plan is, I venture to think, a better way of obtaining the value of colours in terms of standard colours than that adopted by Maxwell. This method gives the values directly, and not by calculation from matches with white. Let us place one slit near each of the extreme ends of the spectrum; that in the red near the red lithium line, and another a little beyond G in the violet of the spectrum, whilst the third slit should be in the exact position in the green, where *the green-blind sees grey.* Now it might be a matter of dispute

as to whether one was entitled to make this last one
of the positions for the slits, for we use it entirely
on the assumption that two of the colour sensations
which we suppose we possess are identical with those of
the green-blind. This might be, or might not be, the
case; but I think it can be shown very easily that the
assumption we are making is more than probably exact.
Having the slits in these positions, we may endeavour
to match the spectrum orange. We mix the red and the
green lights together, and find that the best mixture
is always paler than the orange, but by adding a small
quantity of white to the orange we at once form a
match. In the same way if we have a greenish-blue
to match, we shall find that we can only make the
match when we add a little white to the simple
colour. Now let us shift the position of the slit in
the green just a little—a very little—towards the
blue, and again try to match orange. Do what we
will we cannot find apertures to the slits which will
give us the colour, though it be diluted with white.
It will be too blue or too red, but never exactly orange.
This tells us that there is too much blue in the green
we are using. Next, shift the slit a little towards the
red below our fixed position, and endeavour to match
the blue. We shall find that this, too, becomes imprac-

ticable. The blue is either too green or too violet, telling us that our mixture contains too much green. As the neutral point of the colour blind is the only position for the green slit which enables us to make a good match to both the orange and the blue, it follows that this must be the point where these two colour sensations are so arranged as to be in the proportions required to form white when green is added ; that is, that there is neither an excess of red nor an excess of violet. To come back to our measures of colour. We can make up every spectrum colour with these three colours, and finally divide the luminosity curve into the *colour luminosity.* In all these matches the violet luminosity is very small indeed compared with the red or green. A match with white is now made by a mixture of all these colours, and you will see, from the images of the slits on the screen, that the *luminosity* of the violet is almost a negligible quantity compared with the others. We may, therefore, as a first approximation, divide up the luminosity curve into two parts, one being the luminosity of the green in the different colours and the other of the red. The green, however, is made up of red, of violet, and of an excess of green sensation, which in this case comes practically to a mixture of white with the green sensation. How

can we tell how much is green and how much is white?
Suppose I, as a normal-eyed person, compare the luminosity of the colour coming through the red slit with
that coming through the green slit, and then get the
green-blind to do the same, it is evident that any excess
in the luminosity as measured by myself over that
measured by the green-blind must be due to the green
sensation, and we can also see how
much red and violet make up his
white. We shall not be far wrong,
then, in apportioning the consti-
tuents of the white thus found
between the green and the red;
the violet being, for
the time being, neg-
ligible. We must

Fig. 22.

subtract the red sensation from the green *colour* curve and
add it to the red colour curve : the two curves will then
be very closely the curves of the *red and green sensations.* By causing the green-blind to make mixtures of
red and violet for all the colours of their spectrum, we
can arrive at what must be finally taken away or given
to these curves, though such addition or subtraction of
violet will be small when the luminosities are considered
The accompanying figure (Fig. 22) gives an idea of the

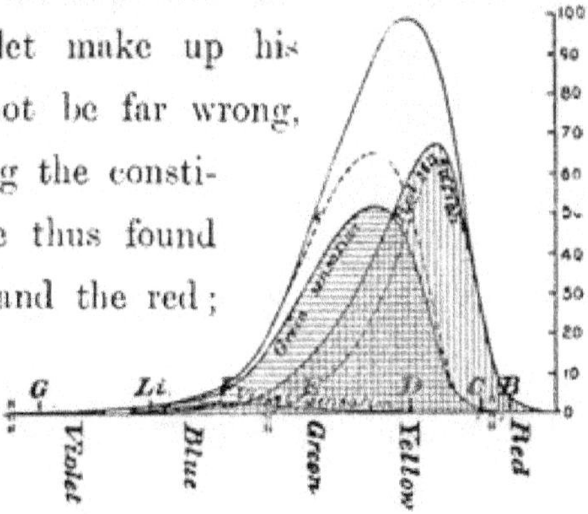

shape and general features of these curves. It may be
remarked that we can check the general accuracy of
the measures of the colour mixtures by calculating or
measuring the areas of the two colour curves, the red
and the green. If accurate, they should bear the same
ratio that the *luminosities* of the two colours bear to
each other (when mixed with a little violet, which is
practically negligible) to form white light. So far, then,
we can utilize the luminosity methods to calculate and
to trace the sensation curves for the normal eye. It
will not escape your notice that the maximum heights
of these two component curves are nowhere near the
parts of the spectrum where the colour is the purest.
Another check to these curves may also be made by
taking the difference in the ordinates of the luminosity
curves of the colour blind and the normal eyed. Too
much stress must not, however, for the moment, be
laid on this, as this method depends on the absolute
correctness of the scale of the ordinates in the curves.
It must be recollected that to the former white light
is deprived of at least one constituent sensation which
is perceived by normal eyes. This, in all probability,
renders the white less luminous to them than those
possessing normal vision, so that the comparisons of
luminosities are referred to different standards.

It may seem a very simple matter to ascertain the correct scale, but it is not, except by the extinction method, which will be described later. At one time General Festing and myself tried to obtain a comparison by finding the limiting illumination at which a book could be read. We got results, but for the purpose in question the values are not conclusive. What we really were measuring was the *acuteness of vision* in different coloured lights. As a good deal depends upon the optical perfection of the eyes under examination, besides the illumination, we must be on our guard, even if there were nothing else against the method, against taking any such measures as being conclusive.

H

CHAPTER VIII.

BEFORE quitting the measurement of luminosity, it may be as well to see whether the curves described are the same whatever the brilliancy of the spectrum may be. We can easily experiment with a very reduced brightness. Upon the screen we have an ordinarily bright spectrum. As the slit, through which the white light forming the spectrum comes, is narrowed, there is an evident change in the relative brightness of the different parts, though the energy of every ray must be proportionally reduced. The red is much more enfeebled than the green, and in brightness the green part of the spectrum looks much more intense than the yellow, which is ordinarily the brightest part. This we have assured ourselves of not only by casual observation, but also by direct measurement. Perhaps I can make this even more decisive to you. Two slits are now in the ordinarily bright spectrum, one in the red, and the other in a green which is near the E line of the solar spectrum. Instead of using one lens to form a single colour patch of mixed light, two parts of a lens, appropriately cut and of the same focal length as the large combining lens, are placed in front of the slits, one bit of lens before each. This artifice enables us to throw the patch of red on one white surface, and the patch of green light on another adjacent to it. By opening or closing one or other of the slits the brightness of the two patches of light are so arranged that there is no manner of doubt but that the red is the brighter of the two. The absolute energies of the rays forming each of the patches are proportionally reduced by closing the slit of the collimator, as before. At one stage both patches appear of about the same intensity. This might be taken for an error in judgment, but to make the change that takes place perfectly plain to you, the rotating sectors are introduced in front of the two slits, and the rays now pass through them. The apertures of the sectors are gradually closed, and we now come to such a reduction that the red is absolutely invisible ; but the green still shines out. It is losing its colour somewhat, and appears of a bluish tint. The reason of this change of hue in the latter we shall shortly see. The sectors are withdrawn and the red re-appears, and is as bright as the green. The slit of the collimator is next opened, and there is no doubt that the red is much brighter than the green, as it was purposely made at the beginning of the experiment. The same class of experiment might have been repeated with the green and violet or the red and violet, and the same kind of results would have been obtained. The violet would have been the last to disappear when the green was so reduced in luminosity that it appeared in the ordinary brilliant spectrum to be equal to the violet ray selected. When the green was of the luminosity given by a slit equal in width to that of the violet, the violet would have disappeared first, owing to its feeble brightness to begin with. Now, if we measure a feebly illuminated spectrum we must adopt some special means to exclude all light, except that of the comparison light and the ray to be measured. This we can do by the box which is shown in the next diagram (Fig. 24).

At one end of a box, shown in plan, is an eye-piece, E. The other end has at its centre a white square of paper of $1\frac{1}{2}$-inch scale. The mono-chromatic beam a, coming from the spectrum through the slit S and the reference beam b of white light,

are reflected from glass mirrors M_1, M_2 to apertures
in opposite sides of the box, and from close to these

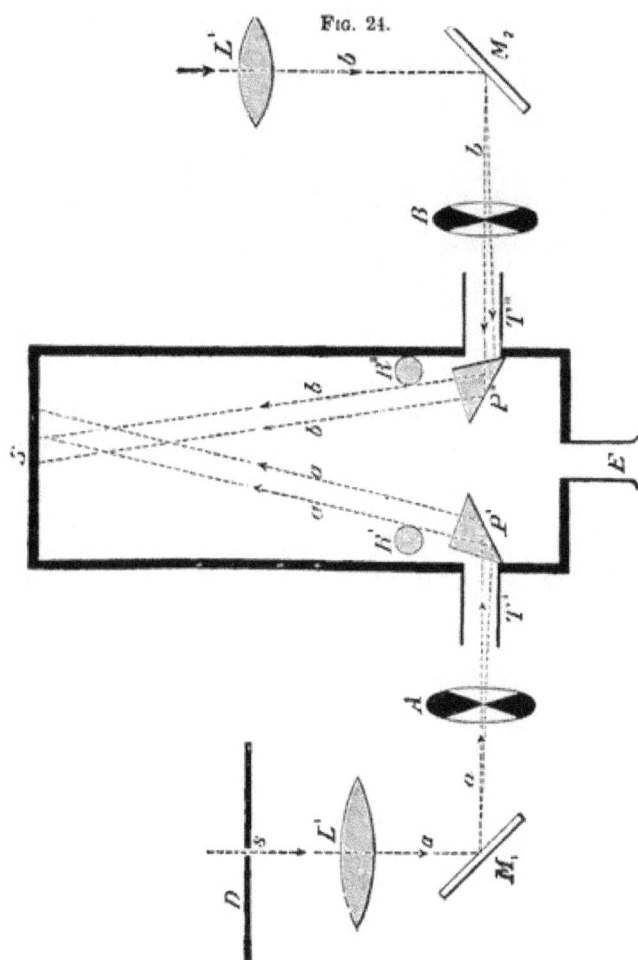

Fig. 24.

apertures by the right-angled prisms P_1 P_2, so as to
fall on and cover S. Rods R_1, R_2' are inserted in the

box in the paths of the beams, so that the opposite
halos of S are illuminated. Diaphragms inside the
box cut off any stray light, and rotating sectors
placed at A and B regulate the intensity of the
beams as required. The sector A is rotated with a
previously determined-on aperture; the white light
coming through B is altered till the luminosity of the
two on the screen, as seen through E, are the same.
Every part of the spectrum can be measured in this
way; the result is shown in the diagram, Fig. 25
(the measures will be found at page 215 in the
appendix). In this case the orange light at D where
it fell on the screen was equal to $\frac{1}{132}$ of an amyl-
acetate light, which, in its turn, is closely ·8 of a
standard candle. In the same figure the luminosity
curve of the ordinary bright spectrum is given for
reference, and it can be seen how the point of
maximum luminosity is shifted into the green, lying
almost over the E line of the solar spectrum. The
maximum, of course, has been made 100 as before, for
had it been drawn to the same scale as the other,
the form of the curve would not have been demonstrated. There is a remarkable resemblance between
it and the curve of luminosity of the monochromatic
vision, and such a resemblance can scarcely be

fortuitous. As a matter of fact, in this we seem to have come to the final curve for low luminosities,

Fig. 25.

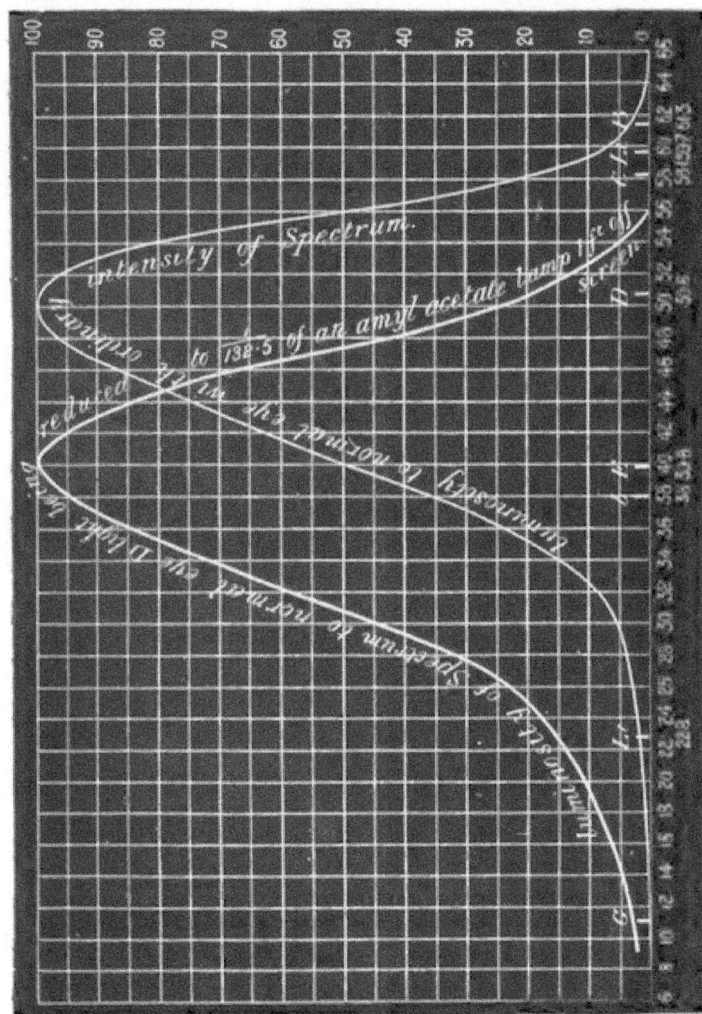

and is almost the same as that observed when the spectrum is reduced to such an extent that it is colourless throughout, a condition that it can assume, as we shall see very shortly. When the spectrum is rather more luminous, it gives a curve of luminosity which is similar to that of the ordinary spectrum when measured by a red-blind person. Here, then, we have an indication that a person with normal vision passes through a stage of red-blindness, as the intensity is diminished before he arrives at absolutely monochromatic vision.

This investigation is of practical as well as theoretical interest, as General Festing and myself quickly discovered when we first made it. The curious colour of a moonlight landscape is entirely accounted for by it. White light becomes greenish-blue as it diminishes in intensity, and the reds and yellows, being reduced or absent, are not reflected by surrounding objects. Hence, moonlight is cold, whilst the sunlight is warm owing to their presence.

When measuring these low luminosities, the various colours will in a great measure disappear. Part of the spectrum will be of that peculiar grey which was shown you in the experiment with the in-candescent light (p. 34). By further experiment it

is possible to arrive at an approximate determination
of the point where all colour vanishes from the
different parts of the spectrum. We use the same
apparatus (Fig. 24) as before, the only difference being
that each of the sectors is movable during rotation.
The apertures of those through which the colour
passes are reduced till all colour on the screen just
disappears, the point being arrived at by a comparison with the white, which is itself also reduced.
The apertures of the first sector alone need be noted,
and from these readings the diagram (Fig. 26) is
made (for measures, see page 216).

This extinction of colour is one which often occurs,
but is seldom noticed. The figure tells us that the
orange is about the last colour of the spectrum
left, some of the others still appearing as greys. The
next to retain its colour is the green, and the most
rapid to lose them are the red and violet. It must
not be supposed that the colours remain of the same
hue up to the time that they vanish. Pure spectrum
red (red sensation) remains the same up to the last,
but the scarlet becomes orange, and the orange
yellower, and the green bluer. This is what would
be predicted from the Young theory if the order of
extinction of sensation be red, green, violet. This

Fig. 26.

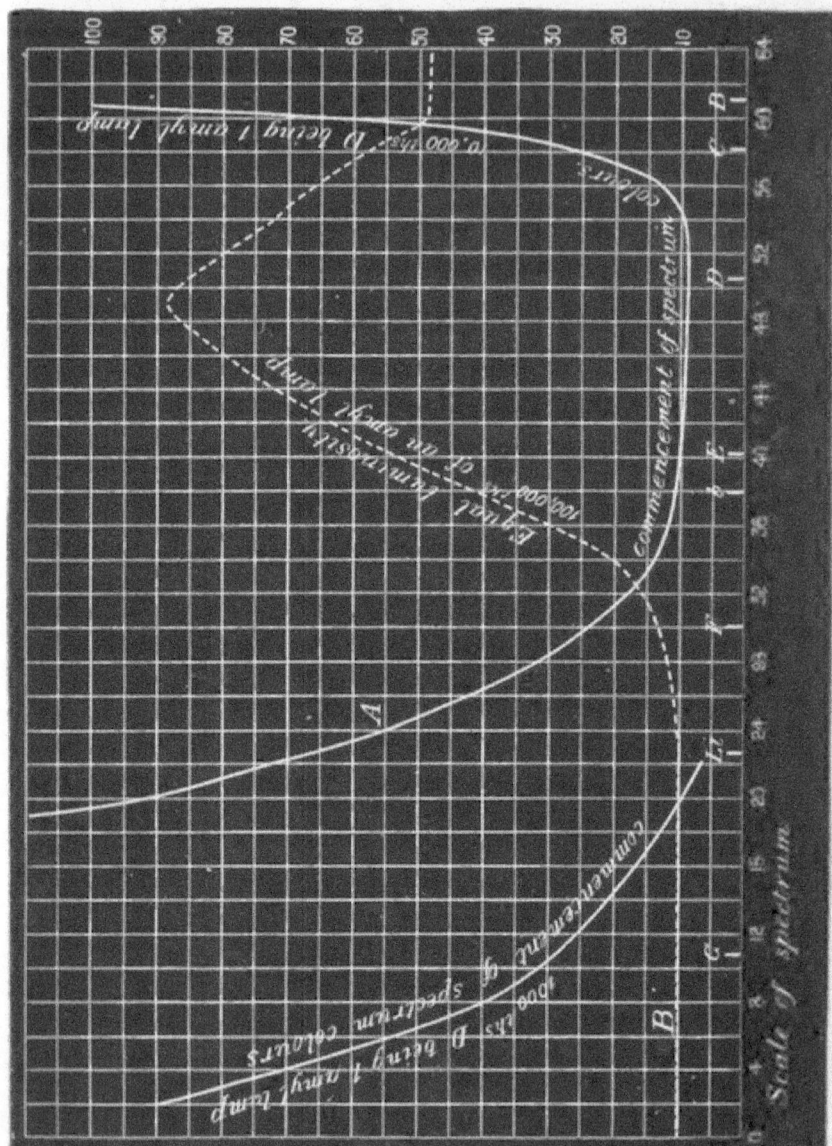

we shall see is the case. At nightfall in the summer the order of disappearance of colour may often be seen ; orange flowers may be plainly visible, yet a red geranium may appear black as night ; the green grass will be grey when the colour of the yellow flowers may yet be just visible. An early morning start in the autumn before daybreak will give an ample opportunity of satisfying oneself as to the order in which colours gradually re-appear as daybreak approaches. Red flowers will be at the outset black, whilst other colours will be visible as grey. As more light comes from the sky the pale yellow and blue flowers will next be distinguished, though the grass may still be a nondescript grey. Then, as the light still increases, every colour will burst out, if not in their full brilliance, yet into their own undoubted hue.

CHAPTER IX.

NOT only, however, may we lose a sense of colour, but we may also lose all sense of light by reducing the energy of the different rays. We have seen that colour goes unequally from the different parts of the spectrum. We may therefore prognosticate that the light itself may disappear more rapidly from some parts than from others. You will scarcely, however, I think, be prepared for the enormous difference which exists in the stages of disappearance of the grey of the reduced red and of that of the reduced green.

But how are we to measure this extinction of light at the different parts of the spectrum? This is a problem which I have attacked during the last few years by a variety of methods; but as is the case with almost every scientific problem, when the mode of attack is reduced to its simplest form, it yields the more readily to solution. If we have a box, like that figured in Fig. 27, and combine it with our

colour patch apparatus, the problem is solved. B B
is a closed box 3 feet long and about 1 foot high
and wide, having two similar apertures $1\frac{1}{2}$ inch in
diameter in the positions shown. The aperture at

the side is covered on the inside
by a piece of glass a, ground on
both sides, and a tube T is inserted, in which diaphragms, D, of
various apertures can be inserted
at pleasure. The most convenient
form of diaphragm is that supplied
with photographic lenses—an iris
diaphragm. E is a tube fitted
at the end
of the box
through
which the
screen S is viewed. S is black
except in the centre, where a
white disc is fastened to it. A
mirror, M, placed as shown, reflects the light scattered
by the ground glass on to the screen S. The rotating
sectors are placed where shown, and are in such a
position that they can be readily adjusted by the
observer. The patch of any desired colour of the

FIG. 27.

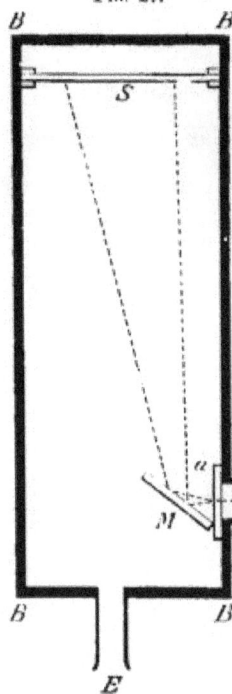

spectrum is thrown on *a*, and an appropriate size
of diaphragm used, so that when the sectors are not
less than 5° to 10° open, the light totally disappears.
We can now make observations throughout the whole
spectrum, and knowing the value of the different
apertures of the diaphragm and the angular opening
of the rotating sectors, we can at once find the
amount of reduction of the particular part of the
spectrum that is being required in order to just
extinguish all traces of light from the white disc
at the end of the box. From these measures
we can readily construct a curve or curves which
will graphically show the reduction given to the
different parts of the spectrum. Fig. 28 gives the
curve of extinction for ordinary normal colour vision.
The spectrum was of such a brilliance that the
intensity of the square patch of light formed on *a*
of the orange light (D) was exactly that of an
amyl-acetate lamp, placed at one foot distance from
the receiving screen. Knowing this, the actual luminosity of all the other rays of the spectrum can be
derived from the curve of luminosity (see Fig. 20).
Extinguishing the various parts of the spectrum
by this plan, it is found that the red rays cease
to stimulate the retina sufficiently to give any

appearance of light long before the green rays are extinguished. It is only the rays in the extreme violet

FIG.23.

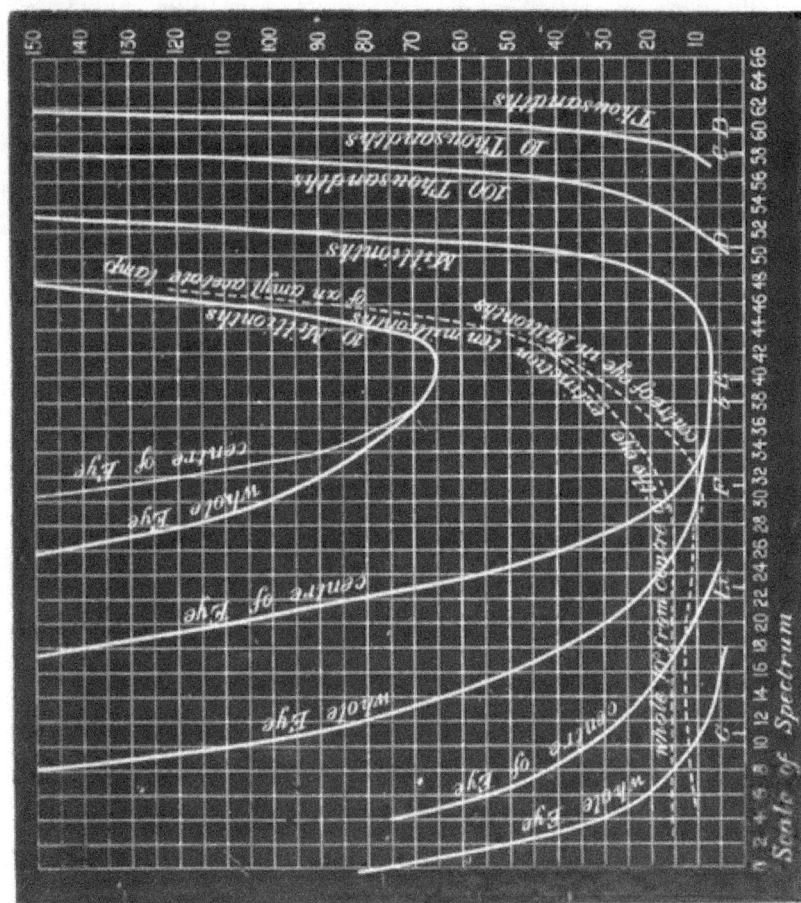

of the spectrum, and which consequently possess
very feeble luminosity, that make any approach

towards requiring the same amount of reduction as the red rays.

There is the fact to remember in making these measures in the extreme red and the extreme violet, that the luminosities of the colours are so small that the illumination of the prism itself, by the white light falling on it, has to be taken into account, since it forms an appreciable portion of the patch of feeble colour. By placing a proper shade of blue or red glass in the front of the collimator slit this white light disappears or becomes negligible, and when the absorption of the coloured glass is known from measurement, we can get a very accurate measure of the extinction of these parts. Some people may propound the idea that the rotating sectors may in such kind of measurements give a false result. Now such a criticism is quite fair, and it is absolutely necessary that it should be answered. Well, to test the accuracy or the reverse of the assumption that such measures are correct, the following small piece of simple apparatus was devised. A and B (Fig. 29) are two mirrors placed at angles of 45° to the angle of incidence of the beam. The path the beam takes can be readily ascertained from the figure. This piece of apparatus was placed in position in front of the spectrum, and the reflected

beams used to form the patches of colour. For
convenience only a small pencil of light was allowed
to issue from the prism, a diaphragm of some ½-inch
in diameter being placed in front of it. This allows
a spot of any desired colour to
fall on the screen, the ground
glass being removed. The slit
through which the spectrum colours
pass is moved along the spectrum,
and a position is arrived at where
the last glimmer of light disappears.

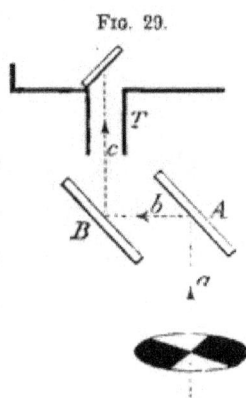

Fig. 29.

The mirrors A and B may both
be of plain glass blackened with smoke on one side,
or one may be plain glass and one silvered, or they
both may be silvered. This, with the power possessed
of altering the aperture of the slit of collimator, puts us
in possession of ample means of making our measures.
We may also use the ground-glass arrangement and use
different diaphragms, which puts a further power of
variation in our hands. I may at once state that the
resulting measurements fell on the curves, obtained by
measurements made with the rotating sectors, a sufficient
proof that the sectors may be used with confidence.
There is still another method which avoids a resort

I

to the sectors. A tapering wedge of black glass can be moved in front of the colour slit, and a different thickness of glass will be required to cause the extinction of each colour. Recently I have modified the extinction box, more particularly for the purpose of using it where the spectrum is to be formed of a feeble light, such as that of an incandescent lamp or a candle. If a really black wedge could be obtained, this would seem to be the best method, but no glass is really black. We have, therefore, to make a preliminary study of the wedge to ascertain accurately the absorption co-efficients for the different rays, a piece of work which requires a good deal of patience, but which, when done, is always at command.

In Fig. 28 two branches of the curves are given at the blue end of the spectrum ; one is shown as the extinction for the centre of the eye, and the other of the whole eye. Of course the former observations were made by looking direct at the spot. This may appear a very easy matter, but it is not really so simple as it sounds. It is curious how little control there is over the absolute direction of the eyes when the light has almost disappeared. The axes of the eyes are often directed to quite a different point. When the extinction for the whole eye is made, the readings are really much easier, as then the eye roams where it likes, and a final disappearance is noted. When the eye has once been invested with a roving commission, it is hard to control it. In making these observations it was therefore advisable to have data for the first branch of the curve, before commencing to observe for the later. The main cause of difference between the two branches of the curve is due to the absorption by the yellow spot.

It might be thought that with the curves (Fig. 28) before us, we have learnt all we can regarding the extinction of light, but is it so? Surely we ought to know something as to the reduction necessary for extinction of the different parts of the spectrum when they are all of equal luminosities and of ordinary brightness.

We arrive at this by simple calculation. Supposing we have two luminosities, *one double the other*, it does not require much thought to find out that you have to reduce the greater luminosity twice as much as the other in order for it to be just extinguished. In other words, if we multiply the extinction by the luminosity, we get what we want. Now, in the curves before us, we have taken the luminosity of the yellow light near

D as one amyl-acetate lamp, and that has a height in
the curve showing the spectrum luminosity very closely
approaching 100. We may, therefore, multiply the
extinctions of a ray by the value of its ordinate
in the luminosity curve and divide the result by
100, and this will give us the extinction of each
colour, supposing it had the luminosity of an amyl-
acetate lamp. A portion of the curve so calculated is
shown in the same diagram (Fig. 28) as a dotted line.
It appears at the violet end as an approximately
horizontal line, and then starts rapidly upwards, and
would, if carried on to the same scale, reach far out
of the diagram ; but at the extreme red it would be
found to bend and again become horizontal. I would
have you notice that the same is true not only for
the extinction observed with the centre of the eye
through the yellow spot, but also for the whole eye.
Such straight, horizontal parts of the curve must mean
something.

In the diagram (Fig. 16) of colour sensations we
see that in each of these two regions there is but one
sensation excited, viz. the violet and the red. Now,
if these sensation curves mean anything, the reduction
necessary to produce the extinction of the same sensation when equally stimulated should prove to be the

FIG. 30.

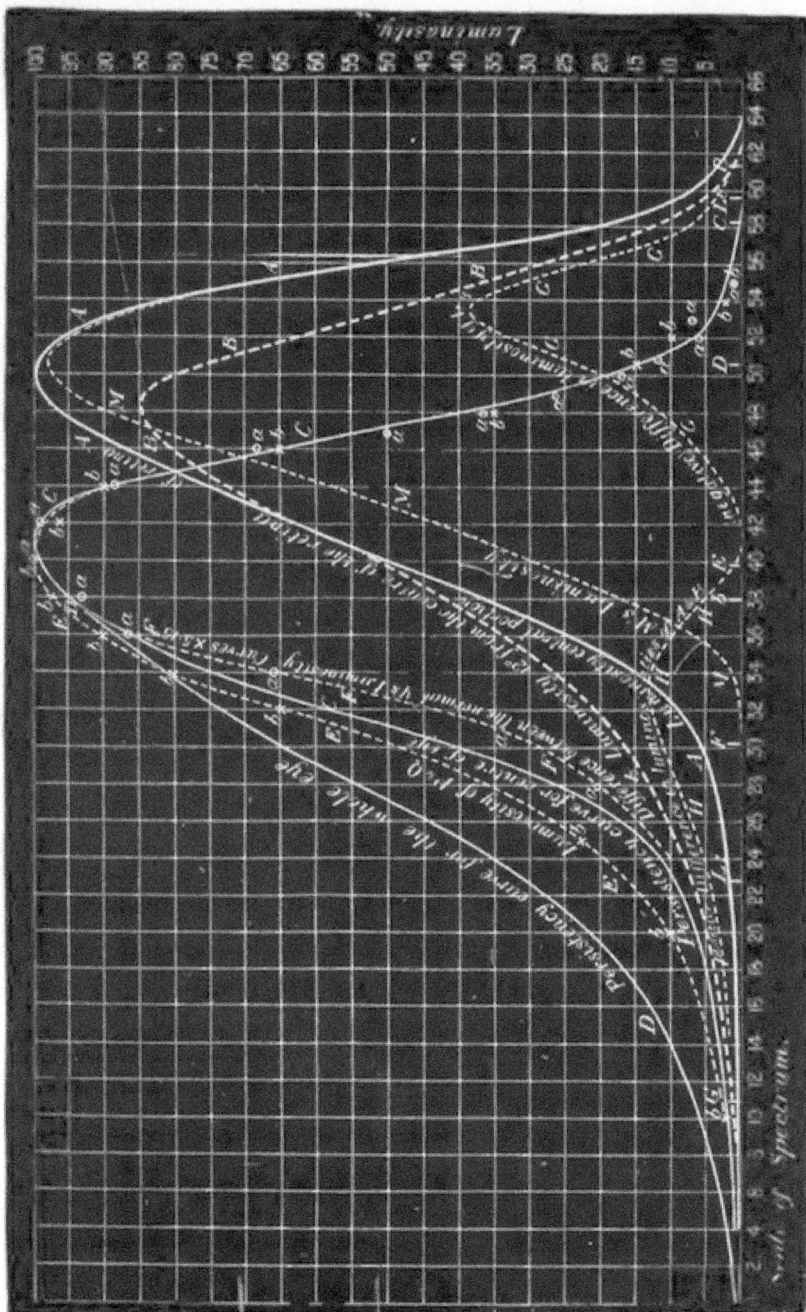

same, for there is no reason to the contrary, but exactly
the reverse. *Primâ facie*, then, taking the Young
theory as correct, we may suppose that these horizontal
parts are due to the extinction of one sensation. Let
us treat it as such, and go back to the original extinction curve shown in the continuous lines. The parts
of the curve which lie over the fairly horizontal dotted
line, at all events, should be the extinction curve
of the same sensation, but more or less stimulated or
excited. As before explained, if we have double the
stimulation at one part of the spectrum to that we
have at another, the reduction of the greater luminosity
to give extinction will be double that of the lesser.
If, then, we take the *reciprocals of the extinction*, it
ought to give us a curve which is of the form of
some colour sensation; and when we arrive at
the maximum, we may for convenience make that
ordinate 100, and reduce the other ordinates proportionally. This has been done in Fig. 30 in the
curves C and D. For the sake of a name my col-
league and myself have named such curves "persistency curves." Perhaps some other name might be
more fitting; but still a poor name is better than
none at all.

When the persistency curve was scrutinized to see

what might be taken as its full signification, I
must confess that the result astonished us somewhat, though we ought not to have been surprised.
The persistency curve C, when applied (in a Euclidean
sense) to the curve of luminosity recorded for the
men who had monochromatic vision, almost exactly
coincided with it. In other words, by far the
largest part of the extinction was due to the extinction of the sensation which in the monochromatic
vision was alone excited. If this be not the case,
there is something in colour vision which no theory
which I am acquainted with can account for. Then,
again, the persistency curve agrees with the curve
of luminosity when the intensity of the spectrum is
very feeble, which is another coincidence of a remarkable character which some theory should explain.
[Fig. 30 gives, besides the persistency curves, the
luminosity curves of the normal eye, of monochromatic vision, and of the violet-blind; and an ex-
aggerated curve of the difference between the normal
luminosity curve and that of the violet-blind, and
others which I think will be found useful for general
reference.]

What sensation is it that is last extinguished, and
which is possessed by a certain class of colour vision?

In the Young theory it can only be the violet sensation. It is certainly not the green, and much less the red. It does not correspond, however, very well with the violet sensation shown in Fig. 16, but more with one which should be in the blue.

In making the extinctions of light, it is quite necessary that certain precautions should be taken to avoid error. All my audience know that when going from bright daylight into a cellar, in which only a glimmer of light is admitted, but little can be seen at first, but that, as the eye "gets accustomed" to the darkness, the surroundings will begin to be seen, and after several minutes what before was blackness comes to be invested with form and detail. So it is with the extinction of light in the apparatus described. Observations carried on before the full sensibility of the eye is attained are of no value. A recorded set of observations will show this. A light of a certain character was thrown on the extinction box, to be extinguished, and the observer entered the darkened room from the full glare of daylight. The eye was placed at the eye end and kept there, and the extinctions were made one after the other till they became very fairly constant. The following is the result :—

Times of Observation.		Readings.
At the commencement	. . .	1·0
After 38 sec.	. . . .	3·2
After 53 sec.	. . .	4·9
After 1 min. 11 sec.	.	6·9
After 1 min. 44 sec.	.	10·5
After 2 min. 43 sec.	. .	17·0
After 3 min. 44 sec.		27·5
After 4 min. 52 sec.	.	43·0
After 5 min. 59 sec.	.	63·0
After 6 min. 41 sec.		78·0
After 7 min. 28 sec.		89·0
After 8 min. 32 sec.		96·0
After 10 min. 46 sec.		103·0
After 12 min.	. . . .	103·0

(For convenience the first reading is unity; the other numbers are the *inverse* of the extinction value.)

The eye apparently, under the conditions in which these observations were made, was at least 100 times more sensitive to very faint light after twelve minutes than it was at the beginning, and that then concordant readings could be made. It will now be quite understood that before any serious measures can be made this interval must elapse, and also that the light, finding its way to the end of the box to illuminate the spot, should never be strong, otherwise the eye might lose its sensitiveness.

CHAPTER X.

BEFORE considering the subject of the extinction of light by other types of colour vision, attention must be called to what has already been brought before you. The various colours of the spectrum have to be reduced to the following amounts before they suffer extinction, the orange light at D being of the value of one candle. (See appendix, page 217, for complete tables.)

	Reduction in Millionths.			Remarks.
B	10,000	or	$\frac{1}{100}$	approximately pure red sensation
C ...	1,100	or	$\frac{1}{909}$	rather more scarlet
D ...	50	or	$\frac{1}{20000}$	orange light
E ...	6·5	or	$\frac{1}{154000}$	a green chosen by Maxwell as a standard colour
F	15·0	or	$\frac{1}{67000}$	beginning of the blue
Blue Lithium	85·0	or	$\frac{1}{11700}$	a good sample of blue
G	300·0	or	$\frac{1}{3300}$	approximately pure sensation of violet.

If we make these same colours all of the luminosity of one amyl-acetate lamp (·8 of a candle), we find that the numbers are as follows :—

			Reduction in Millionths.				Reduction in Millionths.
B	...	...	300	F	...	...	·9
C	...	...	225	Blue Lithium			1·1
D	...	...	48	G	...		1·1
E	...		3·3				

These numbers are remarkable, and we may enforce what they mean in this way. The energy of radiation, and of light also when of ordinary luminosity, varies inversely as the square of the distance from an incandescent body when of small dimensions. But from the above it seems that a white screen receiving the rays from an amyl-acetate lamp in an otherwise perfectly dark place, and having a colour which stimulates the red sensation alone, would be invisible at 58 feet distance, for there would not be enough energy transmitted to stimulate the red perceiving apparatus sufficiently to give the sensation of light. If it were an orange light, such as sodium, of the same luminosity, we should have to move it from the screen 142 feet before the same result was attained. With the green light at E, the distance would be 550 feet, and with the violet the distance would be increased to 1000 feet. The reduction in intensity of white light, which, when of ordinary brightness, is warm, would make it colder, for the red would disappear, and finally the residue of light, just before extinction, would become a cold grey, due to the absence of all colour. The changes in hue that would occur are variable, the variation being due to the loss of colour of the different rays for different amounts of reduction, and then their final extinction. We can place two patches of white light on the screen, and gradually reduce one in intensity, keeping the other of its original value. No one would expect that the two would be dissimilar in hue, as they appear to be when the former is moderately near the extinction value. If we wish to see this perfectly, we should use an extinction box and view it away from the surroundings, which must be more or less slightly illuminated.

It has already been stated that the persistency curve for persons who have normal colour vision is closely the same as that recorded for those who are of the monochromatic type. As this is so, we must expect to find that the persistency curve of these last is the same as their luminosity curve. We put this to the test of experiment and found that our reasoning was correct, for the persistency curve could be almost exactly fitted to it. (See table, pages 217 and 222.) The slight difference between them can be credited to the fact that the whole eye may have been brought into use during the extinction observations, the centre of the eye not being exclusively used. The Figure 31

shows both the extinction and the persistency curves,
and also the curve of luminosity for the normal eye.

Fig. 31.

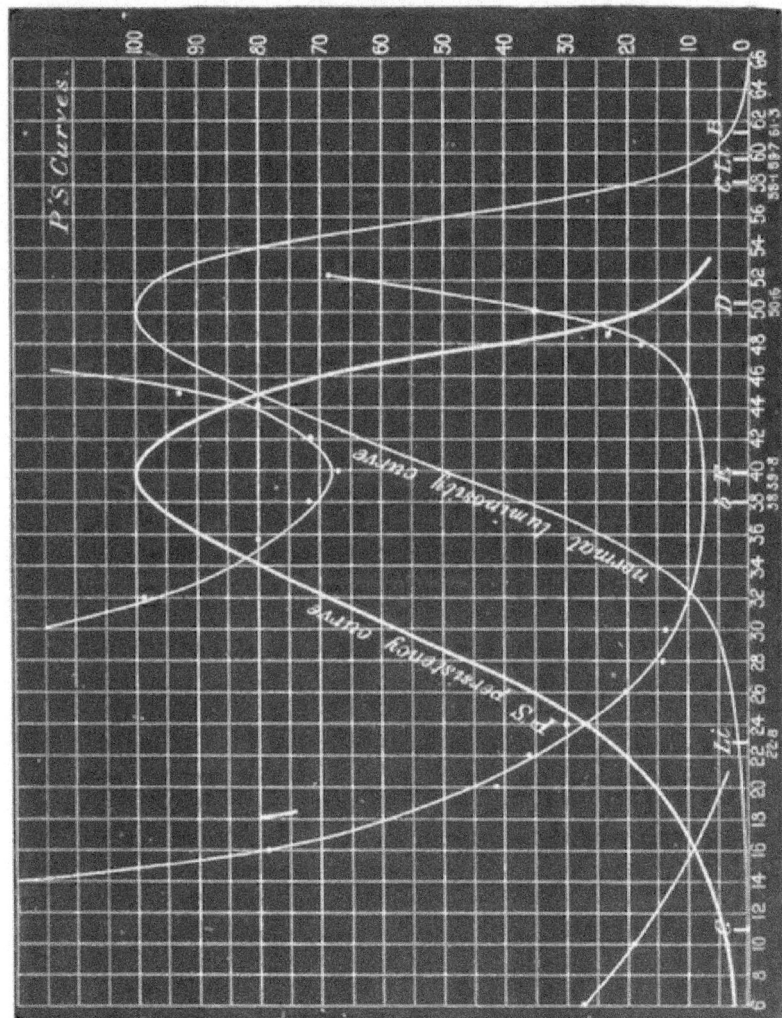

The former were derived from a case P. sent for examination. P. and Q. are brothers, each of whom possesses but one colour sensation, and examination showed that their vision was identical. Mr. Nettleship has kindly given me the following particulars regarding them :—" Their acutes of vision (form vision) in ordinary daylight is only one-tenth of the normal. A younger sister and brother are idiotic and almost totally blind, and in one of these the optic nerves show clear evidence of disease. Hence, the colour blindness of P. and Q. must almost without doubt be considered as the result of disease, perhaps ante-natal, involving some portion of the visual apparatus." A lack of acuteness of vision would be expected from the small amount of light they perceive compared with normal vision. The fact that two of a family, not twins, possess exactly the same colour sense, and that their extinction curves are entirely different to those suffering from post-natal disease, but similar to those of normal vision, point to their colour blindness as falling in the same general category as that of the congenital type. To this I shall refer again.

We may reason still further. With the red- and green-blind the violet sensation is still present, and we

may therefore expect that their extinction curves, and
consequently their persistency curves, should be alike,
and should also agree with that made from your
lecturer's observations. A study of Figures 32 and 33
will tell you that such is practically the case. The
former shows the luminosity, the persistency, and the
extinction curves of a completely red-blind subject, and
the latter the same curves for a green-blind subject (see
pages 223 and 224). Both were excellent observers, and
their examination was easy, owing to the acquaintance
with scientific methods. The accuracy of their results
may be taken as unquestionable. Each of them may be
taken as a representative of their own particular type of
colour blindness. There is an agreement between them
at the violet end, but a deviation at the red end of the
spectrum. The general form of the curves indicates
that the same sensation is extinguished last in all.
Now, have we any other criterion to offer? We have.
In the first instance, we have the violet-blind person
to compare with the others, and also another observer
who had monochromatic vision, but whose sensation
was different to that of the two monochromatic cases we
have so far brought to your notice. We have already
stated the peculiarities in colour nomenclature of the
violet-blind case. His curve of luminosity for the spec-

trum was taken (page 227), and when compared with the curve of normal luminosity, it became evident that in

FIG. 32.

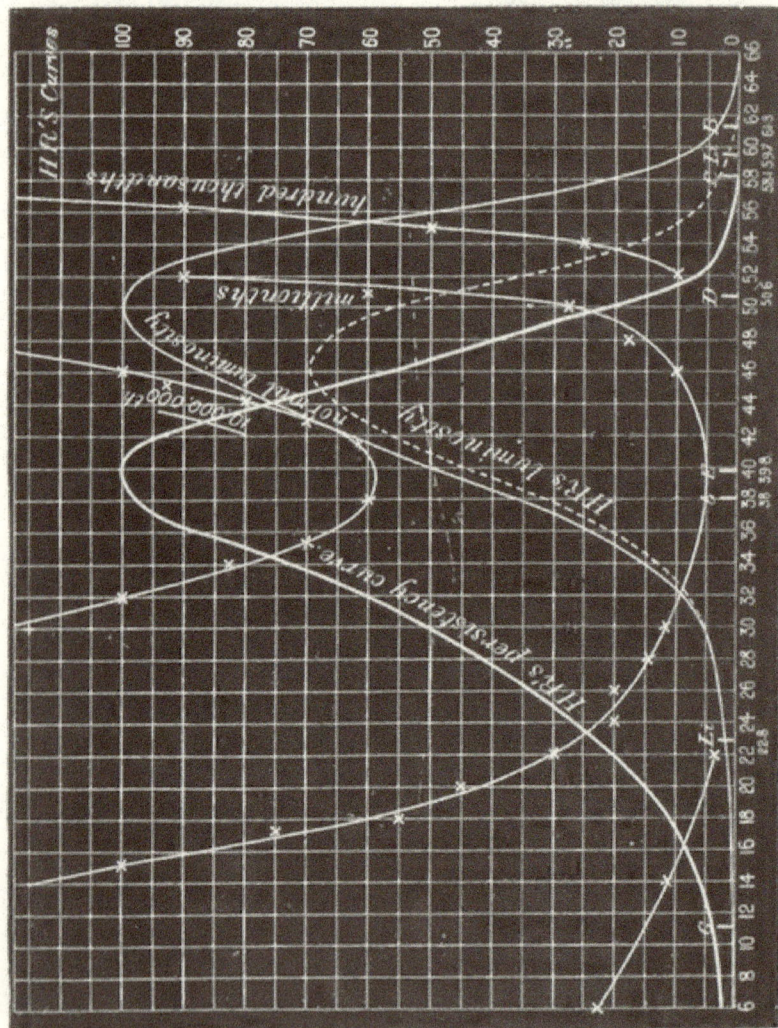

the red and up to the orange his measures were those
which a normal eye would make; but that the luminosity fell off in the green, and finally disappeared
to an immeasurable quantity in the violet (see Fig.
30, curves M and F). If his measures of spectrum
luminosity are deducted from those of the normal eye,
and the ordinates be increased proportionately to
make the maximum difference 100, the figure so
produced, when compared with the *luminosity* curve
obtained from the monochromatic observers, was found
to be the same, and consequently with the persistency
curves above referred to. Endeavours were made to
gain a good extinction curve, but the results were not
as successful as could be desired; but it was ascertained
that, without doubt, his most persistent sensation was
not more than $\frac{1}{180}$ as lasting as that of the normal
eye, or to put it in another way, his green at E was only
extinguished when the energy falling on his eye was
180 times greater than that at which it vanished with
the normal eye. This plainly teaches us that the
missing sensation was that which, when present, is
ordinarily the most persistent.

The next is a case of monochromatic vision, which
differs from those previously brought before you, and I
cannot do better than describe it in the words which

K

General Festing and myself employed in our paper in
the " Philosophical Transactions."

FIG. 33.

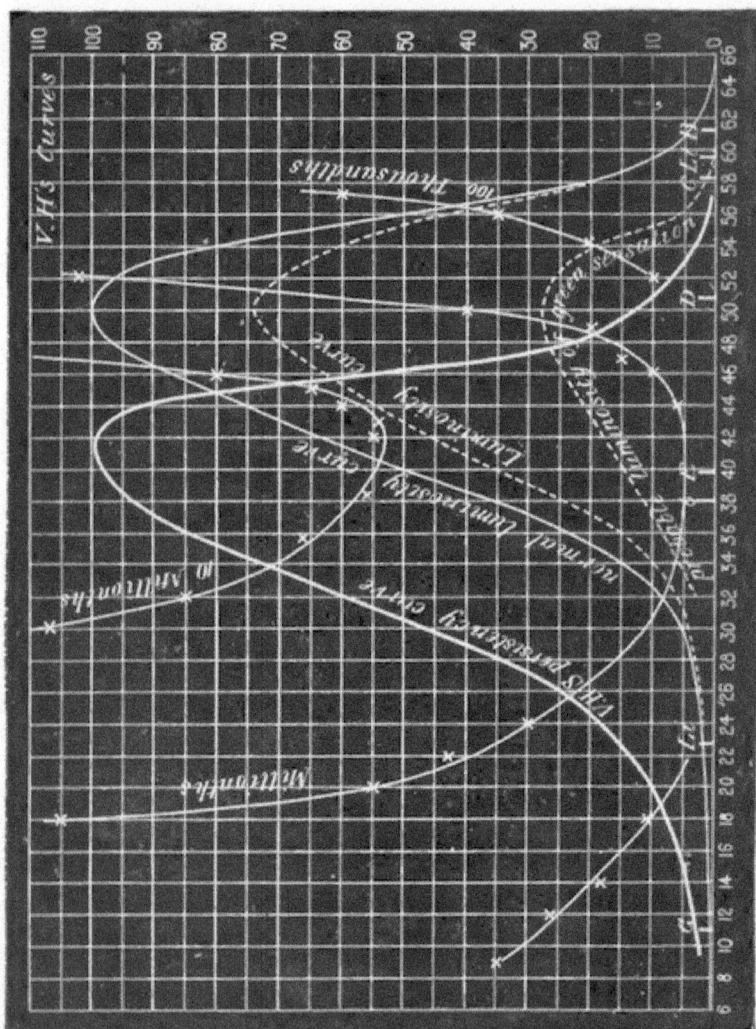

The patient (B. C.) had been examined by Mr. Nettleship, who kindly secured his attendance at South Kensington for the purpose of being examined by the spectrum and other tests. [Mr. Nettleship states that this case is without doubt a genuine case of congenital colour blindness, without any trace whatever of disease.] B. C. is a youth of 19, who has served as an apprentice at sea. His form vision is perfect, and he is not night blind. He can see well at all times, though he states that on a cloudy day his vision seemed to be slightly more acute than in sunshine. He was first requested to make matches with the Holmgren wools in the usual manner, with the result that he was found to possess monochromatic vision. He matched reds, greens, blues, dark yellows, browns, greys, and purples together; and it was a matter of chance if he selected any proper match for any of the test colours. Finally, when pressed, he admitted that the whole of the heap of wools were "blue" to him, any one only differing from another in brightness. The brighter colours he called "dirty" or "pale" blue, terms which eventually proved to be synonymous. We then examined him with patches of monochromatic spectrum colours by means of the colour patch apparatus. He designated every colour as "blue," except a bright yellow, which he called white, but when the luminosity of this colour was reduced he pronounced it a good blue. So with white, as the illumination was decreased, he pronounced it to pass first into dirty blue, and then into a full blue.

Colour discs were then brought into requisition, and it was hard at first to know how to make the necessary alterations, owing to the terms he employed to express the difference which existed between the inner disc and the outer grey ring. By noting that a pale "blue" passed into a pure blue when the amount of white in the outer ring was diminished, and that the inner disc was described as "pale" or "dirty" when the outer ring was described as "a very full blue," we were enabled to make him match accurately a red, a green, and a blue disc separately with mixtures of black and white.

The following are the equations :—

$$360 \text{ red } = 315 \text{ black} + 45 \text{ white.}$$
$$360 \text{ green} = 258 \text{ black} + 102 \text{ white.}$$
$$360 \text{ blue } = 305 \text{ black} + 55 \text{ white.}$$

With these proportions he emphatically stated that all were good blues, and that the inner disc and outer ring were identical in brightness and in colour.

It may be remarked that this is a case of congenital colour blindness, and that there is reason to believe that some of his ancestors were colour blind.

Before using the discs an attempt was made to ascertain the luminosity of the spectrum as it appeared to him. His readings, however, were so erratic that nothing could be made out from these first observations, except to fix the place of maximum luminosity, the terms "pale" and "dirty" puzzling us as to their real meanings. After the experience with the discs we had a clue as to what he wished to express by pale or dirty blue, which only meant that the colour or white was too bright, and on making a second attempt he matched the luminosities of the two shadows as easily as did P. and Q., the other cases of monochromatic vision. The method adopted was to diminish the white light illuminating one shadow to the point at which he pronounced it a good blue, when a slight alteration in the intensity was always sufficient to secure to his eye equality of luminosity between it and the coloured shadow without his perceiving any alteration in the saturation.

The curve of luminosity, Fig. 34, is a very remarkable one, being different in character to that of P. and Q., the maximum being well on the D side of E. A great falling off in the luminosity when compared with that measured by the normal eye will be noticed both in the blue and in the red. (For measures see page 225.) The evidence was therefore presumptive that B. C.'s colour sensation was neither red nor blue, but probably a green.

Fig. 34.

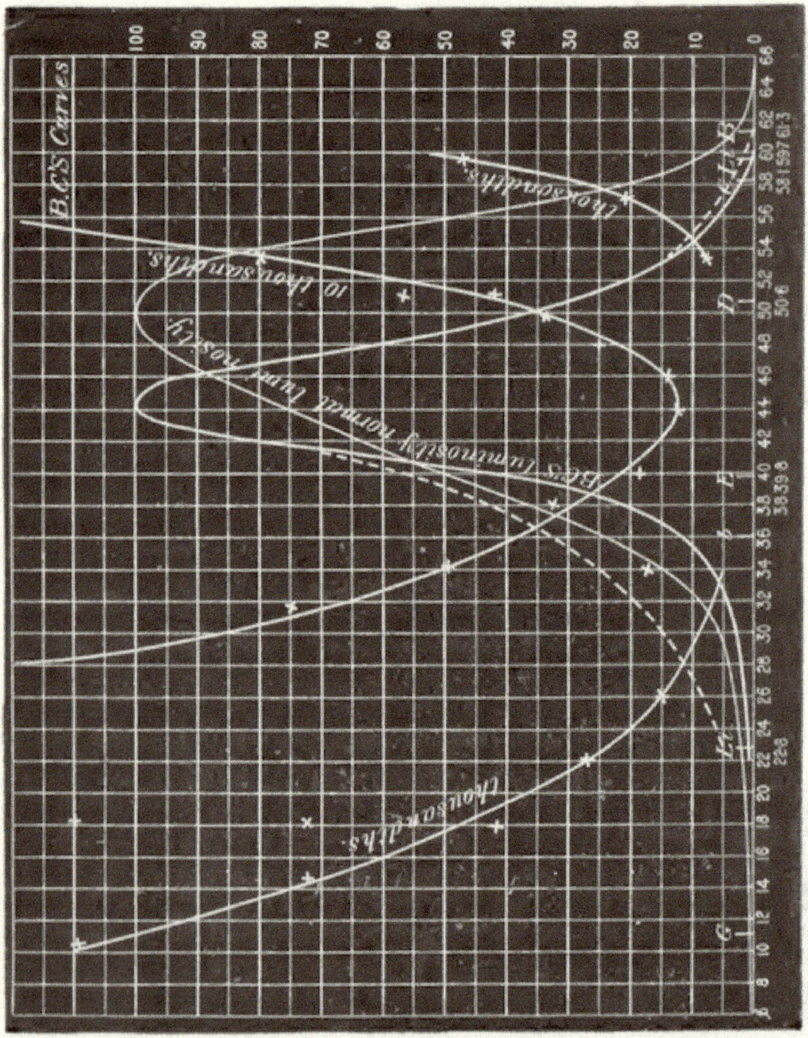

B. C.'s Luminosity and Extinction Curves.

The next test was made to throw light on this
point. He made observations of the extinction of the
different parts of the spectrum. His observations
were very fair, except on the violet side of F, where
they became slightly erratic, but by requesting him to
use all parts of his retina to obtain the last glimpse of
light, a very concordant curve resulted, as shown in Fig.
34. Some of his observations at this part were
evidently made with the centre of the retina, for they
gave readings which, when the "persistency" curve
was calculated, and these observations treated as part of
the extinction, agreed with the luminosity curve. We
may, therefore, conclude that B. C. has a region in the
retina in which there is an absorbing medium corresponding to the yellow spot of the normal eyed. This
is diagrammatically shown in Fig. 34 by the difference
in height of ordinates in the persistency (dotted) and
the luminosity curves. On the red side of the maximum
the two curves are practically identical, except from
Scale number 54. At this point it is probable that the
white light which illuminated the prism vitiated the
readings to some degree. At the violet end something
similar, doubtless, occurs, but it is masked by the
difference that exists in the extinction by the central
part of the retina and that of the whole eye.

It must, however, be remarked that the amount of reduction of the intensity of a ray to produce extinction is very different for B. C. and for the normal eyed, or for the red- and green-blind or for P. and Q. B. C. can bear nearly 200 times less reduction for the rays near E. We have already pointed out that the same is practically the case with M., whom we presume to be violet-blind. We may therefore deduce the fact that the monochromatic vision in this case is of a totally different type to that of P. and Q., and that the last sensation to be lost is the same as that of M. If any violet sensation were present in either, the fact would be made evident by the order of the extinction. The sensation of B. C. is thus apparently the green sensation, though that this particular sensation is exactly the same as that absent in the green-blind is not certain.

The observations made by the different types of the colour blind seem to me to throw great light on the theory of colour vision. They show that when the violet sensation is present, according to the Young theory, the extinction shows its presence; and that where this sensation is absent, the reduction of light necessary to produce extinction is greatly less, and may with great certainty be attributed to a different sensation being the final one to disappear.

CHAPTER XI.

I HAVE so far spoken only of normal, or physiological, colour blindness; a peculiarity, or defect, present at birth, and, as far as is at present known, irremediable, but not associated with any defect of the visual functions, or with any disease or any optical peculiarities. What the nature and seat of this defect may be— whether in the eye or in the sensorium—is at present unknown, although some of the characteristics of the deficiency in colour sensation, I believe, seem to indicate the existence of a special part of the brain endowed with the functions for perceiving colour.

But cases are well known to medical men in which colour vision, normal to start with, fails in greater or less degree in connection with disease. This part of the subject is large and very complex, and requires for its full elucidation an acquaintance with the diseases and disorders of the eye. Many of the phenomena accompanying acquired colour blindness, however, are

of great interest to the physicist in his study of colour
vision, more particularly in regard to the test of the
truth of any particular theory. Through the kindness
of several medical men, and Mr. Nettleship in particular, I have had the opportunity of examining by
the colour apparatus several types of colour blindness
due to disease. One feature, common, I understand,
to all, or nearly all, cases, is the presence of some
disease of the optic nerve. Defective sight—from loss
of transparency of the cornea, the crystalline lens, or
other transparent parts of the eye—does not interfere
with the perception of colour; nor is true colour
blindness, as I am informed, well marked, if present
at all, in disease limited to the choroid and retina (see
Fig. 1). Even in cases of the disease of the optic
nerve, medical authorities tell us that great differences
exist in the amount of colour defect, and that *although
the colour defect always goes along with some other serious
visual loss, either of form, light, or field,* the relation
between these several factors of the visual defect is by
no means always the same, so far as can be judged by
the tests commonly used by ophthalmic surgeons. They
tell us that in some cases of disease of the optic nerve,
colour vision when tested by the wool test, which will
be described shortly, may be almost perfect, whilst the

capacity for reading test letters of the alphabet may be
extremely bad, and *vice versâ*. It seems that in some
cases these discrepancies cannot be accounted for; but
in others the facts can be explained by the limitation
of the disease to certain fibres of the optic nerve.
Thus, if those fibres which supply the yellow spot
region of the retina are alone involved, direct, or
central, vision will be much damaged both for form and
colour, whilst a little further from the centre of the
field, the visual functions in such a case are often quite
normal. From what has been said in the opening
chapters, this will be understood to be that the colour
vision is perfect, but the definition of form more or less
imperfect. We are told that cases of this type have
long been known and are comparatively common, and
often favourable as regards recovery; that the mischief
may affect one optic nerve, or both; that when both
are diseased the malady is usually due to the action of
some toxic substance, and that of all substances known to
have this particular effect on the optic nerves tobacco
is the most important. I dwell a little on this variety
—damage to form and colour sense at the centre of the
visual field of each eye from limited, and usually
curable, disease of the optic nerve—on account of its
interest to myself in the investigations I have made,

and also on account of the degree of practical importance which it assumes in connection with the proper reading of signals and coloured lights. These cases of " tobacco amblyopia," as it is pathologically called, are, of course, always found in men; and it may occasionally happen that such a man, if an engine driver, signalman, or a look-out man on board ship, may still see form sufficiently well to see his signals, but may mistake their true colours. From evidence given before the Committee of the Royal Society on Colour Vision, it appears that the disease causing this type of colour blindness is usually produced by the over-use of tobacco, aided by mental depression and a low state of health. As we have no sumptuary laws, cases of tobacco blindness must frequently occur, and it should be the care of all who have the management of railways or shipping to take measures for preventing persons suffering from this disease from occupying posts which require perfect colour vision in order to prevent the possibility of loss of life.

Congenital colour blindness can at once be discovered, and its possessor be excluded from any post in which normal colour perception is necessary, but with this type a single examination is no safeguard, as it may be developed at any period of a man's career. The disease is, I believe, a progressive one, and at first is most generally unrecognised, the deficiencies of vision being usually slight at its commencement. It is very often brought to the notice of the sufferer by finding he is unable to read. The words at first seem only slightly indistinct, but later become undecipherable, and as time goes on he is unable to even see the letters. He or his friends then usually think it time to consult the specialist. In tobacco amblyopia the area of insensibility is central, and it may subtend a very small angle or one which covers a considerable portion of the field. I am not aware that it ever extends over it all, but it very generally covers the yellow spot. Now as the eye naturally receives the image on the centre of the retina, it follows that, as the ability to distinguish some colours is absent in that particular region. the patient is practically colour blind, though he can distinguish them on most parts of the retina which are not affected. As regards form vision, it was mentioned in the first chapter that in a healthy eye it is much more acute at the centre than towards the periphery, and instances were given of the angular distances apart that black dots on a white ground were required to be placed to allow their being seen as separate objects when the images were received on the centre of the retina, and at the periphery. Sharp definition may be said to be almost confined to 3° of angular distance at the centre, and most probably this is a happy state of affairs, for if we could see equally distinctly with the whole field of vision, the mind would be distracted from the object which it wished primarily to contemplate.

Bearing in mind the want of definition beyond 3°, and the indistinctness caused by a diseased central area, it will not be surprising to find that form vision in these cases is imperfect throughout, though the colour perception outside such area may be unimpaired. But, practically, men suffering from this disease are colour blind to coloured objects, such as a signal light on a railway or a ship's light at sea. They may see that there is light at the distant signal or on the bow of a vessel, but will be unable to interpret correctly the colour. The colours which fail to make visual impressions are the reds and greens. Some will distinguish yellow, and very nearly all will distinguish blue with the centre of the eye. If a bright spectrum be thrown on the screen, and a tobacco-blind person be requested to name the colours of the different parts pointed out to him, it is often the case that as his eyes follow the pointer he will tell you that in the extreme red he sees no light, but in the bright red he sees dull white. The bright yellow he will tell you is a pale yellow or white, according as his case is a moderate or bad one; the green he will call white, and the blue and violet he will designate correctly. At the same time that his eye is turned away to another colour, he will see the true colour of the part of the spectrum which he has just incorrectly named, but it will disappear again as he turns his eyes back again. This tells us that his sense of colour is apparently unaffected outside the diseased area.

At page 10, a description has been given of the manner in which the field for colour and light has been determined, and if this same method be pursued with persons suffering from this form of colour blindness we get some remarkable results. Fig. 35 is the chart of the eye for red and for white, which was made by a case of tobacco blindness. The yellow spot is entirely affected, and, as is very common, it extends to the blind spot in the eye. At no place within that area can red be seen, though blue is immediately recognised. The extent of the field for white is that found under normal conditions, and except for the diseased area the same is true for the red. The fields for both eyes are given: that for the left eye in the left-hand chart, and that for the right eye in the right-hand chart. The small dark spots within the 5° area are places where the colour sensation is most defective. The part in the central dark area shaded with lines in this direction ////shows the portion of the field which is insensitive to red, though not to *light*, whilst the remainder of the shaded central area indicates the extent of the field which

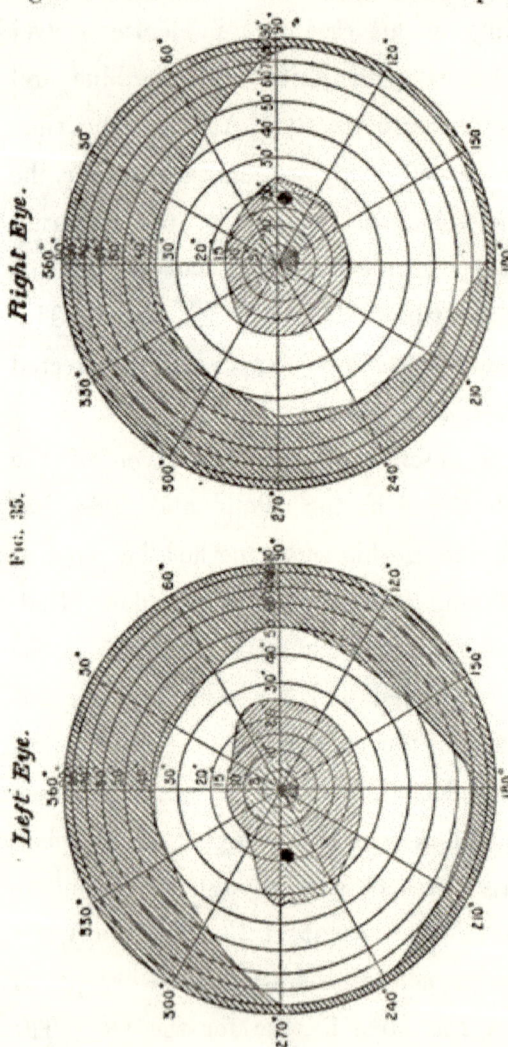

Right Eye.

Left Eye.

Fig. 35.

is sensitive to red. The field for light generally is also shown by the (approximately) rectangular unshaded area. Although the area occupied by the insensitive part of the retina is small compared with the whole, yet it is in that part which is used for distinct vision.

For testing for colour the apparatus, Fig. 3, arranged so that the patch of colour has the white patch alongside, is the most useful, but it is as well then to use a surface of patch about $\frac{1}{2}$ inch square only, and thus to confine the image as nearly as may be to the spot on the retina which is defective. These cases of central scotoma are by no means very easy to test; for it frequently happens that before they are able to distinguish that there are two patches side by side, they have to approach very close to the screen. If this be the case, however, it will usually be found that the patches of $\frac{1}{2}$ inch side are still efficient, as the near approach of the eyes to the screen indicates a wide area as being affected, so that the image still lies within the diseased retinal area. In some instances the colours named will vary very considerably; sometimes, for instance, a red will be named as grey, and then immediately after as pale red. This is generally due to the diseased area being small, and a very slight change in the direction of the axis of the eye causes

I.

it to be seen in nearly its true colour, part being
viewed with the diseased and part with the healthy
portion of the retina. With the wool test, which we
shall describe later, it is the commonest thing possible
for colour-blind persons who have a central scotoma
to match accurately the different test-skeins, for the
reason that the images of the skeins of wool are so
large that they are received on the parts of the
retina which are not diseased. These same colours,
however, if presented to them in small patches, will
inevitably show the defect in vision.

With this end in view, I have had a set of brick-clay
pellets some $\frac{3}{16}$-inch in diameter, painted with water-
colours mixed with soluble glass solution of the same
colours as the wools. These are placed in a shallow
tray, and presented to patients affected with this
central colour blindness to pick out all the pellets
which match reds and greens. They will tell you
that they see neither one nor the other, though they
will pick out the blue pellets unerringly. A red pellet
they will match with a red, green, grey, or a brown
one, and a green one with the same. If, however,
you instruct them to direct their eyes a few degrees
away from the tray, they will tell you they see all
the colours, and as they endeavour to pick them out,

they, with a natural instinct, direct their eyes again
to the collection, when once more the colours vanish.
It is almost piteous sometimes to see the distress
which this simple test occasions. The sight of the
colours for an instant and their immediate disappearance in the cases that I have tried, seem indicative
of something terrible, for they usually have no idea
of the cause of this (to them almost miraculous)
phenomenon. I have seen these colour blind tested
with a pair of ordinary bull's-eye lanterns, placed side
by side, with diaphragms of moderate size with
coloured glasses, which can be changed at will, in
front. At twelve feet distance they will often see both
lights as one, but as they approach they will make out
two lights and call them both white, or sometimes
they will make a guess and call a green red, or *vice
versâ*. It goes without saying that such eyesight is
useless for reading signals, and indeed for any
purpose whatever. Sometimes, but I believe this is
rare, no colour whatever can be distinguished.

CHAPTER XII.

I WILL now give in full the result of the examination of a patient who was suffering from tobacco blindness. X., aged thirty-six, a commercial traveller, was suffering from rather severe tobacco amblyopia. The scotoma was a very marked one, and the loss of colour sensation most complete. Mr. Nettleship, who furnished the case, has kindly added the following remarks on the case :—

His acuteness of vision was $\frac{6}{36}$ with the right eye and $\frac{6}{26}$ with the left. He smoked half-an-ounce of "shag" daily and drank about four pints of beer. His sight had been failing for about two months. As is common in early stages of this disease the ophthalmo-scope revealed no decided changes at the optic discs.

He passed the test of the Holmgren wools satis-factorily, proving that the usual vision was normal for colour, but failed at once with the pellet test.

The objects in view were to test his perception of the spectrum colours, and then the extent of his retinal

field for colour. This last is not recorded here. The
spectrum colours were reduced to uniform luminosity

Fig. 36.

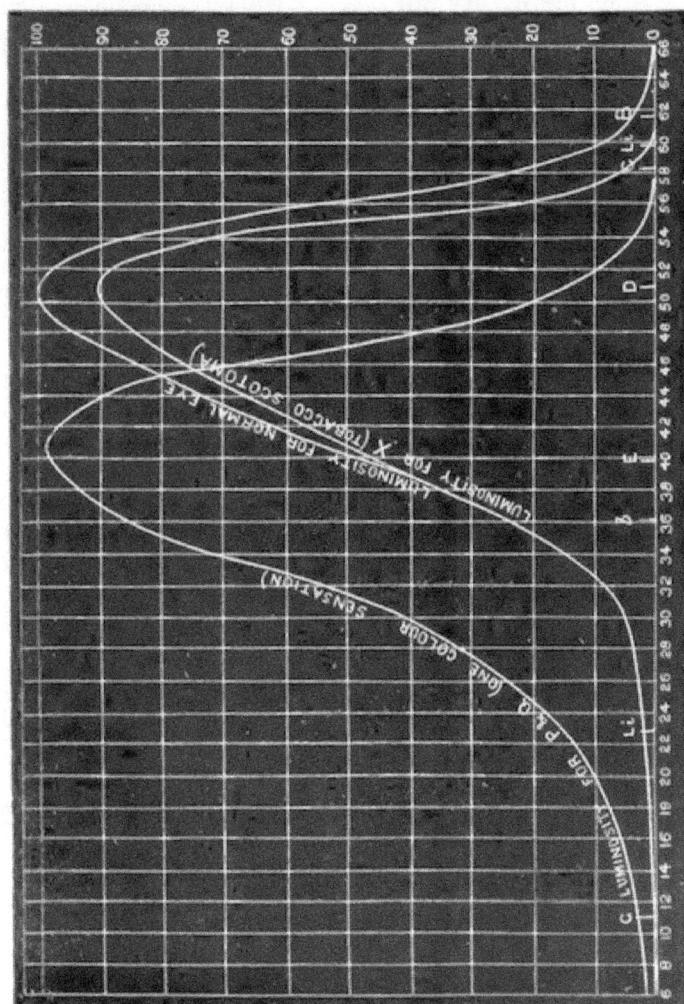

between λ 4600 and λ 6600. Diaphragms containing holes of different sizes were placed in front of the last prism, and thus a round spot of monochromatic light of the same luminosity was produced upon the screen when a slit was passed through the spectrum. From the red end to λ 5270 he called the whole of the colours white, and from that point he began to see blue, called the colours bluish and blue. When the full illumination for all the colours was used, the same results were obtained. From this examination it would appear that he was totally deprived of the sensation of any colour except of blue. A subsequent examination of his perception of the luminosity of different rays, however, has to be taken into account, for in the first examination he had no light of pure white with which to compare the colours. In the next experiments, a strip of white light was placed in juxtaposition to the colour, and the results were slightly different. The table below gives his luminosity measures (Fig. 36). Col. I. is the empyric scale number, II. is the wave-length, III. the luminosity of the colour to the normal eye, IV. the luminosity to X., and V. the ratios of III. to IV.

In the diagram, his luminosity curve X. is shown, its area being 1400 against 1650 for the normal eye. His central perception of light, as arrived at by the extinction method, was only two-thirds of that of the normal eye ; hence his area of luminosity should be 1100. As it is 1400, the ordinates of the above curve should be multiplied by 0·8, to compare with that of the normal eye.

I.	II.	III.	IV.	V.	Colours to X.	Spectrum colour to normal eye.
Scale No.	Wave-length.	Luminosity to the normal eye.	Luminosity to X.	IV./III.		
60	6730	7·3	0	0	Sees only the white stripe	Red.
57	6423	32	10	0·31	Calls red yellowish, and white bluish	Scarlet.
55	6242	65	38	0·65	,, ,,	
53	6074	96	86	0·89	Both one colour	Red-orange.
51	5920	99	90	0·91	,, ,,	Orange-yellow.
47	5660	92	83	0·90	Calls green a little blue; white he sees as white	Greenish-yellow.
43	5430	69	625	0·90	,, ,,	Yellowish-green.
40	5270	50	46	0·92	,, ,,	Green.
32	4910	8·5	9	1·06	Sees blue as blue, and white yellowish	Greenish-blue.
31	4960	7	8	1·14	,, ,,	Blue.
26	4680	3	3	1·00	,, ,,	Blue.

His readings of luminosity were made without any hesitation, and were concordant for each observation, which is not to be wondered at, as the matches, except at the blue end, were practically matches of different mixtures of black and white.

It appears that the white which X. sees as white is the same as the orange sodium light, and that the red he sees is yellowish. The mixture of this yellowish-white with the blue makes white. He sees a little blue in the spectrum colour at λ 5720, so it must be taken that at that point of the spectrum he begins to see colour—a point which is considerably lower than that given by his preliminary examination of the spectrum colour, and due, no doubt, to the fact that in this experiment he had the white light of the positive pole of the electric light to compare with it. It seems probable that what X. called yellowish was really a sensation of white mixed with a very small quantity of red sensation, for he saw no yellow in the orange, in which that colour would be most easily distinguished on account of its luminosity. Red light, when strongly diluted with white light, to the normal eye is often called orange.

As, practically speaking, the colour vision of X. is confined to blue and white, it is of interest to note the difference in luminosity at the different parts of the spectrum that is registered by him and by P., who had blue (violet) monochromatic vision. To facilitate the

comparison, the luminosity curve of the latter is shown
in the diagram.

FIG. 37.

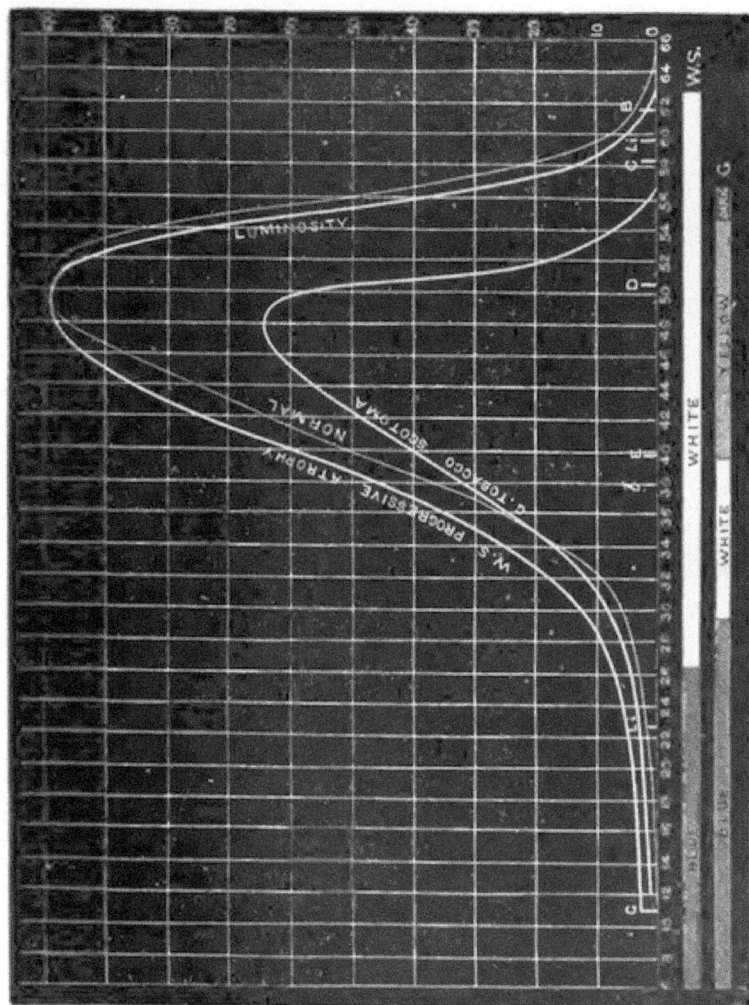

The thin line curve is the normal curve.

Perhaps another case of a patient suffering from tobacco blindness may be quoted, as it will show the differences that exist in recognising the colours of the spectrum, and that the shorter the visible limit of the

TABLE OF LUMINOSITY FOR G. *See* page 153.

Scale No.	Wave-length.	Reading.	Colours named by G.	Colour of spectrum to the normal eye.
57	6423	0		Scarlet.
55	6242	3	No colour	
53	6074	11	Colour " yellow," white " blue "	Red-orange.
51	5919	34	,, ,, ,, ,,	Orange-yellow.
50	5850	60	,, ,, ,, ,,	
49	5783	64	Colour " gold," white " sky-blue "	Yellow.
45	5538	59		
40	5270	40	Both white	Green.
35	5042	18	,,	
30	4848	10	,,	
29	4807	6	Colour " very pale blue," white as white	Blue.
26	4707	4	Colour " blue," white " white "	
20	4518	3	,, ,, ,, ,,	
10	4248	2	,, ,, ,, ,,	Violet.

spectrum at the red end, the more pronounced is the extent of the colour blindness. G. suffered from a very well-marked tobacco scotoma, occupying a consider-able area. His curve of luminosity of the spectrum is shown in Fig. 37. The horizontal band beneath will show the colours which the spectrum colours appeared to match.

G. was tested for light sense by the extinction method, and it appears that the final sensitiveness to light at the central part of the eye was nearly 12 times less than a person possessing normal sense. I may mention that I have examined one, if not two cases in which the patient was not only tobacco blind, but also congenitally colour blind. Though interesting for record, they need not be given in full here.

With these specimens of examination I must leave the cases of tobacco blindness. Although very important, they by no means constitute the sole cases of colour deficiency due to disease. I will give as an instance a case of loss of colour sensation due to progressive atrophy of both eyes which was examined, with Mr. Nettleship's aid. When tested with spectrum colours—a patch of white light being placed in juxtaposition with the colour—it was found that W. S. was absolutely blind to colour from 26·75 (λ 4733) on the scale of the spectrum to the termination of the red of his spectrum, which was close to 63 on the scale (λ 7082). Above scale No. 26·75 W. S. saw blue, and his spectrum was continued normally in the violet. His luminosity curve (Fig. 37) was made without any difficulty, and,

156   *Colour Vision.*

compared with my own, shows a slight deficiency in brightness from the red to the yellow, but his perception of luminosity increases as the blue is approached.

TABLE OF LUMINOSITY FOR W. S. *See* page 155.

Scale No.	Wave-length.	Reading.	Spectrum colours named by W. S.	Spectrum colours to normal eye.
60	6728	3·4	Grey	Scarlet.
58	6520	15·0	,,	
56	6330	41·0	,,	
55	6242	43	,,	
54	6152	69		
52	5996	94		
50	5850	100	,,	Orange.
48	5720	96		
45	5538	88		
42	5373	74		
40	5270	61·5	,,	Green.
38	5172	45		
35	5042	30		
30	4848	12	,,	Blue.
25	4675	6	Bluish	
20	4518	4		
15	4376	3	Blue	Violet.
10	4248	2·5	,,	

He was subsequently tested with colour discs—Ultramarine (U), Red-royal (R), Emerald-green (G), Chrome-yellow (Y), White (W), and Black (B).

It was found that—

165 (U) + 48 (R) + 147 (G) = 75 (W) + 285 (B).

The black reflected 3·4 of white; hence the true equation is—

(i). 165 (U) + 48 (R) + 147 (G) = 84·7 (W) + 275 (B).

(ii). 120 (U) + 240 (Y) = 196 (W) + 164 (B) (corrected)—

> With 260 (U) + 100 (Y) he sees blue.
>
> 250 (U) + 110 (Y)   „   light-blue.
>
> 242 (U) + 118 (Y)   „   no blue.

This last in connection with (ii) shows that his blue perception is neutralised by the yellow, although the yellow to him was matched with white.

I have already shown you a chart of the insensitive area of the retina found in a tobacco-blind case, and it may be advisable that you should see an example of the curtailment that exists, both for light and colour, in the field of vision of eyes in which there is progressive atrophy of the optic nerves. The large black area shows the part of the field that was encroached upon. The dark spots show small areas which are also insensitive. The field for colour shown by the inner shaded area is also encroached upon, and practically the patient was blind in a great part of his field. His form vision was also very bad, and his colour perception feeble. The three charts given in these lectures were brought by Mr. Nettleship, for the information of the Colour Vision Committee of the Royal Society, and by his permission they are re-

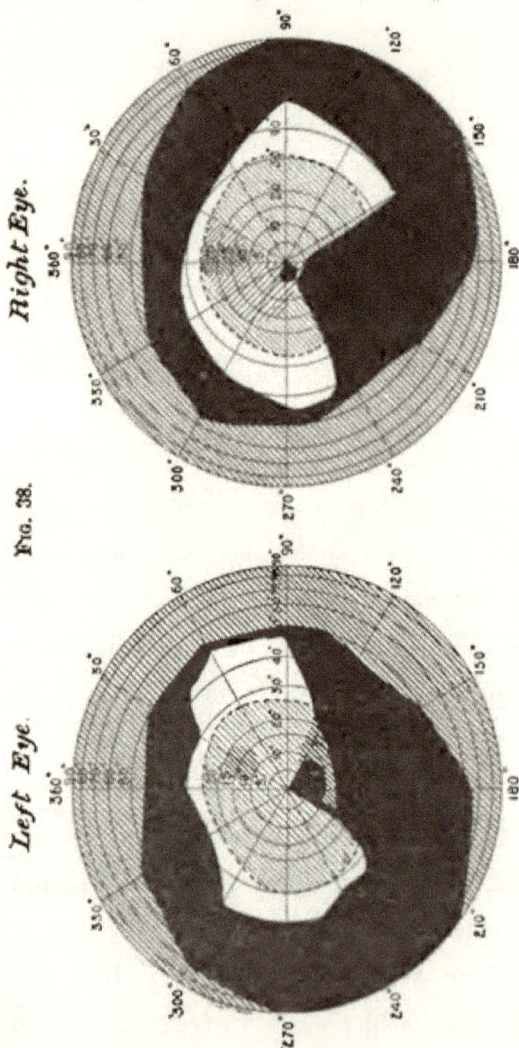

produced here.

Two other cases I may give in some detail, one in which the sensation of colour is totally absent in the left eye, the right eye being normal; and the other in which there is colour blindness of a very rare character. The first case is that of a lady, whom we will call Miss W. It appears from

the history of this lady that she had a slight stroke
of paralysis which affected her left side, and that she
subsequently found her left eye was deprived of all
sensation of colour. It is said by the specialists who
examined her retina that this is a case of atrophy
of the optic nerve. She had very little difficulty
in matching the most brilliant spectrum colours with
the white patch of light. Her curve of luminosity
is given in Fig. 39 (see table, page 228). At 19 of
the scale, which is well in the blue, she had very
little sense of light, though her extinction curve
shows that it extended to some distance beyond. The
eye in which normal vision existed was, during the
examination of the defective eye, bound up with a
handkerchief, and when occasionally she was allowed
to use both eyes, her astonishment was great to see the
colours which she had matched with the white. The
curve of luminosity taken with her right eye coincided
with my own, which throughout we have taken as
normal. From her extinction curve we gather that
there was a marked diminution of sensitiveness to light
in her left eye compared with that of normal vision.
Apparently, in that eye she only has $\frac{1}{25}$ of the normal
sensitiveness to light near E in the green, but her
extinction curve takes the same general form as that

of the normal eye.   The difference between the sets of
ordinates of the two indicates the difference in sensitiveness for each part of the spectrum.

FIG. 39.

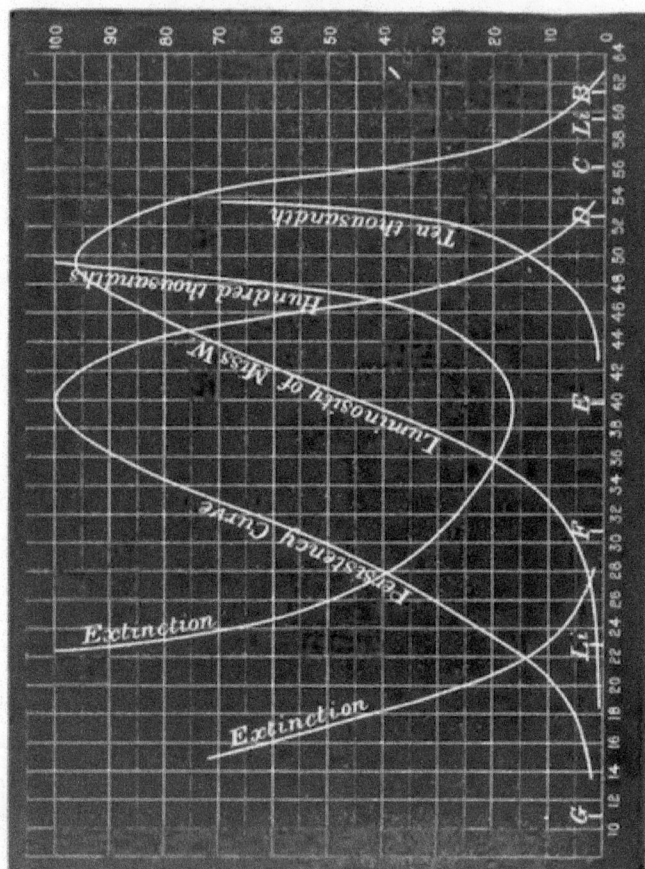

Her persistency curve as calculated occupies the same
position and is of about the same dimensions, when the

maximum is made 100, as that of the normal eye, as it is therefore of red- and green-blind, and also of

FIG. 40.

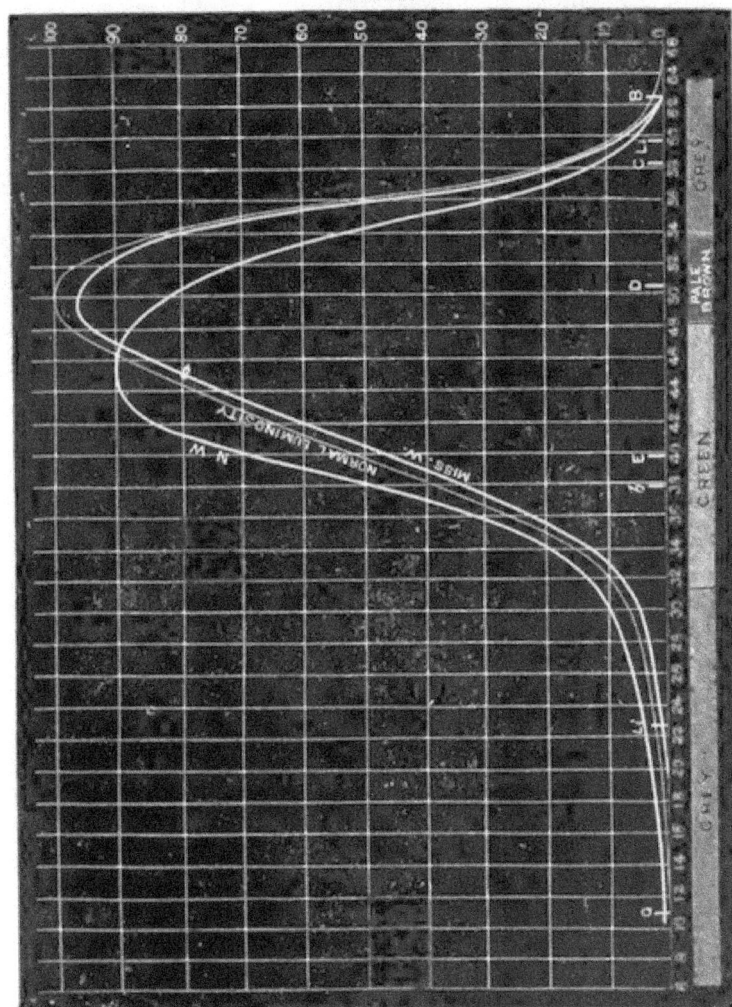

The thin line curve is the curve of luminosity for the normal eye.

the two cases of monochromatic vision. We have in Miss W. a type of colour blindness which no present theory of colour vision accounts for without straining; and it would probably have to refer it to the seat of sensation rather than to the retina alone.

The second is a case of congenital colour blindness and with no trace of disease, brought by Mr. Nettleship to the same Committee. He found that this lady, N. W., mistook blue for red, and it was with some curiosity that this case was examined. Her first examination was as to colour sense with the spectrum colours, a patch of monochromatic light being placed in juxtaposition with an equal patch of white light. At 62·5 (λ 6890) of the scale the light of the spectrum disappeared. As the slit moved along the spectrum, and the white was approximately reduced to equal luminosity, she described all the red as grey, and of the same colour as the white until 53·5 (λ 6110). At this point she said the colour was brownish compared with the white, and this hue continued to her till 48 on the scale (λ 5720), when she said the colour was "neither brown nor green, but both." From 48 on the scale she described the colour as green, when it changed quite suddenly at 31·5 (λ 4905). From this point and in the blue she again began to see grey;

Scale No.	Wave-length.	Reading.	Colours named by N. W.	Spectrum colours to normal vision.
60	6728	3	Both grey	Red.
58	6520	10	,,	
56	6330	30	,,	
54	6152	52	Colour " brownish," white " grey "	
52	5996	70	,, ,, ,,	
50	5850	81	,, ,, ,,	Orange.
48	5720	87	Colour " brownish - green," white " grey "	
46	5596	90	Colour " green," white " grey "	
44	5481	88	,, ,,	
42	5373	82	,, ,,	
40	5270	62·5	,, ,,	Green.
38	5172	46	,, ,,	
35	5042	23	,, ,,	
32	4924	12·5	,, ,,	
31	4886	10	Colour " brownish - grey," white " brownish-green."	
30·5	4862	8·5	,, ,, ,,	Blue.
25	4675	5	,, ,, ,,	
20	4518	3	,, ,, ,,	
15	4376	2·5	,, ,, ,,	
10	4248	1·5	,, ,, ,,	Violet.
0	4010	0·2	,, ,, ,,	

the grey at this end of the spectrum, and also of the white patch, she called brownish-grey. This name must evidently have been a mental distinction, as she described the red end and the white as grey only, and not brownish-grey; and, indeed, she was tested again over that part of the spectrum, and adhered to the previous naming. It would appear to be due to low luminosity, which made the grey appear to her what she called brownish, rather than to any actual difference in hue.

Her curve of luminosity in the spectrum was next taken, and her readings are given in the table above. The curve is shown in Fig. 40. The shaded band beneath it applies to her curve. Miss W.'s luminosity curve is also repeated in the same figure for the sake of comparison.

An endeavour was made to form a series of colour equations with her eyesight by placing three slits in different parts of the spectrum, but without success, although a match with white was made in two positions. One slit was in the orange-red (52 of the scale), another at E, and the third at G; mixtures were made which she said matched the white, but they were so erratic that it was useless to measure the apertures. When the slit in the violet was covered up, a white patch being alongside as a comparison, she called the mixture of red and green "brownish-green"; when the slit in the red was covered she called the mixed light of green and violet "green"; and when the green slit was covered up she called the purple colour a "different kind of brown."

When the first slit was moved into the red near the lithium line she called the colours "green," whenever the green slit was uncovered. A piece of red glass was placed in the white reflected beam, forming a red patch, and a patch of the blue at scale No. 30·5 (λ 4862) was placed alongside, and she matched them in luminosity and in colour. (The dominant colour of the signal glass in question was λ 6220.) She finally was tested with colour discs.

To make white she required

$$130 \ G + 113 \ R + 117 \ U = 72 \ W + 218 \ B.$$

She was then tried with the blue and green discs alone and made a match—

$$258 \ U + 102 \ G = 65 \ W + 295 \ B.$$

An attempt was made to match with the green and red discs alone, but this failed.

She matched the red disc alone with black and white, and also the blue disc alone—

$$360 \ R = 56 \ W + 304 \ B \ \text{(corrected)},$$
$$360 \ U = 60 \ W + 300 \ B \ \text{(corrected)}.$$

With any proportion of R and U mixed together she matched a grey of approximately the same intensity as above, as it might be supposed she would from the last two equations.

Taking the intensity curve of the light reflected from the red disc, it was found to contain a great deal of the part of the spectrum which she called brownish, viz., from $33 \cdot 5$ to 48 on the scale, whereas the blue reflected a trifle of this portion of the spectrum, as did also the green; and this may account for her making a match to grey of U and G, and not of R and G, but it is hard to see why she matched U alone and also R with the grey.

Reviewing the case, it seems that any perception of colour is very small, and that the sensations are green and much less red. From the equations it also seems that she would have matched green with white and black alone, and that $360 \, G = 75 \, W + 285 \, B$. Perhaps the explanation of the matches and names of colours may be that a proportion of colour may be mixed with another without being perceived, but this colour so hidden has still the capability of neutralising a certain quantity of the complementary colour.

CHAPTER XIII.

You have been taken through much experimental work, and possibly it may be thought that there has been too much of it; but now that we are coming to the more practical part of the subject, it will become apparent that a good working hypothesis is absolutely necessary before effectual tests for colour vision can be carried out, and that the reasons for its adoption should be given in full. The question of colour blindness is one of very practical importance, as in certain occupations it is essential that colours should be accurately and quickly known, and that no guess-work should be allowed. Lives have without doubt been lost by a want of proper knowledge of colours, both at sea and on railways. The evidence that such is the case is, as a rule, it is true, merely negative, though there are cases extant where great losses which have occurred can be traced to a deficiency in colour perception. If there be no proper system of tests for ascertaining the defects of signal or look-out men in their colour sense, it is palpable that positive evidence cannot be forthcoming, and this is very much the state of things which exists up to the present time. We hear of collisions at sea and vessels foundering in consequence of the rule of the road not being followed, but at the investigations which follow we have no record that the question of colour perception of the look-out man has been gone into, though there may be conflicting evidence as to whether a red light or a green light was shown. That danger from colour blindness is incurred has for some time been recognised by the Board of Trade, as it insists that all officers of the Mercantile Marine must be tested for their sense of colour, and that their certificates must be endorsed as having failed to pass the colour test should they do so. For my own part, I think endorsement of their certificate is quite inadequate, for it is still open for shipowners to employ them (of course at their own risk). A rejection for colour vision should entail a withholding of the certificate altogether; for it surely is as dangerous that a signal should be misread as it is that the logarithm of the sine of an angle should be misunderstood. If a candidate fails in theoretical navigation, he is not allowed a certificate, but if he only fails in a very practical part of his examination, his certificate is merely endorsed.

The system employed by this department *was* a defective one, and we know of many instances in which candidates have passed the colour test, though they ought to have been rejected, and are at present in the service. The subject of testing for colour vision was brought prominently forward some two or three years ago, and a Committee of the Royal Society, to which I acted as secretary, was requested to consider the methods at that time in force on the railways and in the mercantile marine, and to find one which was not open to objection. It recommended the system that had been elaborated by Holmgren, a Swedish physicist, and known as Holmgren's test, which has long been in force in Sweden and elsewhere. This system has, I am glad to say, been adopted by the Board of Trade, and by most of the railway companies in the United Kingdom. There have been numerous indications that this change of method was necessary. Only within the last month (April, 1894), for instance, I was informed by the Medical Officer who had to examine the employés on a certain railway in Scotland by the Holmgren test that he had found some, amongst others an engine-driver, who were colour blind, and pre-

sumably unfit for the posts they occupied owing to this
defect.

There is one popular objection which is always made
against this test, or indeed against any proper test,
viz., that the examination is not made under the
same conditions which absolutely exist, nor with the
very lights which the candidates have to distinguish
from one another—that is, the red and green lights.
Let me beg of you to remark, that as a mere matter
of guessing, the chances are equal that a man would
name the light shown correctly. If you turn a man's
back to the light, and if he has a coin in his pocket and
deliberately calls heads red and tails green, he will have
a good chance of passing the test; for, if he guessed
rightly three or four times, no one would fail to pass
him on his answers. The great point in a test is to
cause the candidate to *do something* to show that he
appreciates colour. It is this *doing something* and
saying nothing which is the important feature in the
Holmgren test. A man may be ignorant of the names
of colours—colour ignorant it is called—but he cannot
be ignorant of the colours themselves if he has normal
colour vision. As a matter of fact, the colour blind
may possibly distinguish between red and green lights
by having carefully noted, under ordinary conditions

of atmosphere, their different brightness, and by their difference in saturation with their neutral colour. If external conditions are altered, as they are in actual daily life, these slight indications vanish, and the quick naming of the colour to be read becomes a mere matter of chance. A proper test should include all variations that can occur in these respects. It cannot be too strongly impressed upon every one that a man who is colour blind to colours in ordinary daylight is equally so in lamplight, although some shades of colour which are well distinguishable by daylight may disappear when the artificial light is used as the source of illumination.

Now, on what scientific principles should a colour test be founded? We must hark back to a theory for a moment, and as it has been shown that for all essential purposes that of Young answers, we will use it as a good working hypothesis, and it was from this theory that Holmgren himself reasoned. The red- or green-blind see a grey in a part of their spectrum, which to us who possess normal colour vision is green. If then we present such a green to them, they would match it with a grey. If, however, we have a yellowish-green, which is pure green mixed with red, the complete green-blind will not see the green in it, but only the red. The colour to him would be very pale red, and as he sees all such greens and yellows and reds as red more or less saturated, that is, more or less mixed with his neutral colour, any one of these he would match with a green. The red-blind, on the other hand, would see all these colours as green, and he too might make similar matches with them. Suppose now we have a pink skein : the green-blind would see it as a white or bluish-white, for a purple is white to him, and he would match with it either greys or colours having a slight excess of blue in them ; for a green is to him a neutral colour. The red-blind, on the other hand, would see but little green in the pink ; blue would predominate, so he would choose mauves or blues amongst other matches.

Acting on these principles, Holmgren selected his test colours. He chose wools as the most convenient for handling, and also because they present the same colour without sheen when looked at in any direction. His first test colour is a very pale green which contained no blue. Its paleness is a point in its favour. The colour is quite distinguishable by us normal-visioned persons, but it might appear as grey to the red- and green-blind ; for as we who possess normal vision may mix a small percentage of colour with our neutral colour (white) without it being perceived, so may they with theirs (white and green). As the green, when it is to us saturated, would be nearly neutral coloured to them, the very diluted colour which we see in the skein would to them be masked by the addition of white. In any case, if any colour be visible to them, it must be on the red side of the neutral points. A candidate is given this skein of wool, and from a heap of over a hundred skeins, of varying degrees of saturation, amongst which are drabs, yellows, yellow-greens, blue-greens, purples, pinks, greys, and so on, he is asked to select others which appear to him to be of the same colour as the test-skein, though they may be darker or lighter. He will, if colour blind, select some of the colours already indicated. The second test-skein is a pink, which is a purple diluted with white, but much less so than the green, to which it is nearly a complementary in daylight. The candidate is required to select colours which match this, and according to his selections is he pronounced as having normal colour vision or as being colour defective (either completely or partially) to the red or to the green. The case of violet blindness is not important in reading the signals ordinarily used, and therefore in this test no special test-skein is employed. Let us consider what colour we should use. The neutral colour to this form of colour blindness is yellow. If, therefore, we pick out a pale yellow skein, the candidate would pick out greys to match it; or if we gave him the pink skein to match, since he has no blue (violet) sensation, he would match it with a pure red or with a purple.

Where monochromatic vision is under examination, all skeins would be matched with one another indiscriminately—blues, reds, greens, greys will all be a match, some lighter and some darker than the test-skein. I have been told by some who have carried out examinations for colour blindness that this matching is by no means so uncommon as is often imagined. In future it is hoped that most of those who make these matches may be examined by the spectrum test, as it may turn out that a proportion of them will be most valuable theoretical cases.

In making an examination with the Holmgren test, it is almost unnecessary that the candidate should take up a skein out of the heap of wools to form a preliminary diagnosis. The colour blind will not at once pick out an evident match, but will hesitate and evince a desire to appear very accurate in his choice. This indicates at once that there is something amiss. He probably will pick up a skein of the right colour, place it against the test-skein, lay it down and again take it up. Or he will pick up a skein which is evidently incorrect and do the same thing, but perhaps he will return it to the heap and take up another which is equally bad.

He will fumble over making his matches, and eventually have a heap by him which will at once tell the examiner that he is colour defective. I may as well give you an idea of the colours which the colour blind will pick out by a simple experiment. The heap of wools is on the table, and in the pure white of the electric arc light, which is thrown on it from the lantern, every colour is distinct in hue and in intensity. On one side are placed the two important test-skeins, the pale green and the pink. There can be no doubt but that in that heap of wools there are a large number which can be matched with each of them. The red-blind, be it recollected, sees no red, and if I can place in front of the lens of the lantern some medium which cuts off the red as completely as possible, the audience as well as myself will see the colours approximately as the red-blind would do. Such a medium is found in the same blue-green glass that is used for signals on most railways and on board ship. The green-blind, on the other hand, see no green, and if

a medium can be found which when placed in the path
of the light allows no green to pass, the colours in the
heap being deprived of the green would be such as
would very nearly be the same as this type of colour
blind would see.   This glass is covered with a film of
collodion in which fuschin and blue have been dissolved.
It transmits a fine purple and should answer our purpose.   That these two media are what we require
can be readily demonstrated by placing them in front
of the slit of the collimator of our colour apparatus and
throwing the spectrum on the screen.   The spectrum
of white light is now on the screen, and when we place
the blue-green glass in front of the slit, we see that the
red is very nearly entirely extinguished, whilst if we
substitute for it the dyed collodionized glass the green
is absent.   Now, placing the first glass in front of the
lantern lens and switching on the current, the wools are
illuminated with the bluish-green light.   The green test-
skein appears green, and we can proceed to make our
matches, picking out the colours which appear the same,
but taking no heed as to their lightness or darkness.
A dozen skeins are now picked out, and I think the
audience will agree with me that the matches as viewed
in the green light are accurate.   The glass is now withdrawn, and the ordinary white light falls upon the skeins

in my hand. They are a strangely variegated lot as now seen ; we have green shades, yellows, and browns, and greys. Such a variety would tell me that I was colour deficient, but would not be, perhaps, decisive as to what was the exact character of the deficiency. For if the pink glass is placed in front of the lantern you will find the same matches, with one or two exceptions, might have been made. The blue-green glass is once more placed in the beam, and this time I match the pink skein with the wools. A certain number are picked out, and the audience will agree with me that the matches are fair ones. When, however, the glass is withdrawn from the light and we see what colours have been selected, we find that they consist of pale blues, mauves, pinks of various shades, and cerise, and violet. The red in the pink did not affect my eyes any more than would it the red-blind. I am evidently then in this light red-blind, for if the pink glass replaces the blue-green, the matches are impossible. While this coloured light is illuminating the heap I will make matches again. When made, the white light is again thrown on the selected skeins, and this time we have bluish-green and neutral tint together with pinks. The reason of this is evident, there is no green visible ; the bluish-green contains besides blue a certain amount of yellow,

N

which, in its turn, contains red, and the grey must be
pink. To the green-blind, for reasons already given,
the blue-green looks white, as does the pink, and therefore the two are matched together. The grey is also
degraded white to him, and therefore he also matches
that with them. The matches which the violet-blind
would make can be well exemplified by placing in the
beam of light a yellow glass, or a glass coated with
collodion in which "brilliant yellow" has been dissolved. By this plan, then, we can in some measure
produce the effect of colour blindness on ourselves,
and very interesting it is to compare theory with
the results obtained in this manner. There is no necessity to have recourse to the electric light for this
purpose. If matches are made with such media held
before the eyes in ordinary daylight, the same results
will be obtained. I have often examined through these
same media the matches made by the colour blind, and
been able at once to settle the nature of the
defective vision from which they were suffering.
It must be remembered that the colours transmitted
through these two glasses are not *absolutely like* the
whites which the two classes of colour blind see respectively, though they approach it.

We can imitate even more exactly the matches that

they would make by matching white light with a mixture of red, green, and violet of the proper hues, and covering up the red or green slit, and then placing the test-skein and the matches in the colour so formed. From the other skeins viewed in the same light can be picked out the matches which would be possible. There is very little chance, if any, of a mistake about them being made when this plan is adopted.

CHAPTER XIV.

HOLMGREN's test, although a qualitative one, is most accurate in allowing a diagnosis to be formed, but it sometimes happens that a candidate is not satisfied that he has failed in passing the test, and wishes for another examination. Such a re-examination is best carried out by the spectrum method, which I will now describe.

The test with the spectrum is a very decisive one, and can be carried out with the patch apparatus (Fig. 3), page 19. Personally, I like to have some idea of the kind of colour blindness, if any, which exists by first using the Holmgren test. Should these tests show that a candidate is colour blind in any degree, a very excellent beginning is to try and find his neutral point in the spectrum—if he has one. To arrive at it we place two patches of light on the screen, one of colour and the other of white, the rotating sectors being in the last-named beam, and ask him to say when the two colours appear alike. It must be remembered that white is coloured from the effect of contrast as long as the colour alongside differs from it. A good *point de depart* is with the slit in the yellow, then to move it into the red, and then gradually to push it into the green. When here, if colour blind, he will say, "The two patches are nearly alike, but that the white is rather pink or green," as the slit gets further towards the blue. The operator, whilst changing the colour, alters the sectors so that the luminosities are about the same. A point will be reached when the colour blind will say, "Now they are both alike, but one is rather darker than the other." The sectors are altered until he says they are both alike, and the observation is satisfactory when he declares the two patches of light are both alike in colour and in darkness. It is curious how misleading the word brightness is to some people who are uneducated. I find it much safer to ask which is the darker colour, rather than which is the brighter. A little patience will always enable you to get a good observation. The place in the spectrum which is the neutral point is now noted. The neutral point is again found, but this time commencing in the blue. The same procedure is adopted as before, and we thus get a second reading for it, and the two will be found to be very close to one another. In difficult cases, four or five observations may be made, and the mean taken as a close approximation. So far the spectrum test has not shown whether the observer is red- or green-blind, except by comparing the position of the neutral point with that usually found by the two types. We have, however, an unerring criterion by the luminosity method. The red is placed beside the white, and he is asked to say which he considers the darker; he will give an answer of some kind, and probably protest that the two colours are not alike. A soothing answer will disarm his objection, and he will quickly see what you mean. If he be red-blind he will match in brightness a brilliant red and a feeble white; if he be green-blind he will make a match very similar to normal vision. In the case of the red-blind the slit is then moved into the extreme red, when he will say he sees but one patch of light, whilst the green-blind will see it as a person of normal vision would do. If time permits, the whole luminosity curve may be taken and registered. This is not essential, but interesting for reference. Where complete colour blindness exists, it should be possible to cause him to match a green with a red. To do this a second instrument, as described in page 18, may be used, but it is quite sufficient if a piece of red glass, such as is used for railway signals, or of

bottle-green glass, be placed in the white beam. There
is then a red or green patch alongside the patch of
spectrum colour. The red will stimulate the red sensation of the green-blind, but not being spectrum red it
contains a certain amount of yellow, which stimulates
the green sensation if the observer be red-blind. The
green is of such a colour that it will stimulate both
the red and the green sensations. In the path of the
reflected beam between G and the prisms (Fig. 3) a
sheet of plain glass is inserted, which reflects a proportion of white on to the red patch. The sectors are placed
in this beam. If the red glass is being used, the slit is
moved into the green near E, and the colour blind will
say that both are the same colour, but one darker than
the other. By opening or closing the slit in the spectrum, he will possibly say that both colours are alike
and of the same darkness, but he may say one is paler
than the other, in which case the white light must be
increased or diminished by means of the sectors till
equality of tone is established. This applies to the
red-blind and the green-blind. The former will require
a very bright red to match a feeble green, whilst with
the latter the red will require a fairly light green.
When the green glass is used the spectrum colour patch
should be red, and the match be made as before. With

the violet-blind the neutral point will be in the yellow, and with monochromatic vision matches can be made throughout the spectrum. So far it will be seen that no mention of any colour is required. It may next be advisable to ask him the names of colours. This is best done by placing the white patch of light over the spectrum colour patch, and opening and closing, as may be required, the sectors. If the sectors are closed it is very probable that correct guesses may be made, for then the colours will be saturated, and the colour blind, if they are intelligent, will know that a green to them is white or pale in colour compared with red, though of the same hue. If white be mixed with the red the wrong name is bound to be given, for they will be unable to distinguish it from the green, because it is then a less saturated colour. Passing from green to red and mixing the colour more or less with white, the most—I was going to say grotesque—telling mistakes are made. A further excellent test is to place a cell containing a solution of bichromate in the path of the reflected beam, and cause the observer to match its colour with the light coming through two slits, one in the red near C, and the other in the green near E. Defective colour perception will be well demonstrated. There are various other artifices which can be employed in the spectrum test, which would be too long to re-count here, and if there be two sets of apparatus the tests are practically unlimited in number.

There are cases in which an observer who may have normal vision may wish to be reported as colour blind. A seaman's life is not always a happy one, and a boy on a training-ship, knowing that a failure in colour vision will free him from a sea life, may be anxious to be told he has failed in colour vision. By "coaching" in the Holmgren test he might manage to obtain a "failure," but a malingerer is sure to be detected by the spectrum method of testing. He may call diluted red green, and he may declare he sees a neutral point in the spectrum, but if he be tested with the diluted colours near his supposed neutral point, he is sure to fall into a trap. He will make a mistake in calling a patch green when it ought to be white, or white when it ought to be green, if he were truly colour deficient—indeed, a malingerer has no chance of escaping detection with the spectrum tests. It is not an uninteresting experiment to get an acute observer who has normal colour vision, and is accus-tomed to the spectrum test, to feign colour blindness, and examine him in this manner. He never fails to make such mistakes as would lead to his detection.

With the partially colour blind the same procedure may be adopted. In examination by the Holmgren wool test, slight mistakes will be made in matching the first two test-skeins. With the spectrum test the red will require a greater dilution with white before it will be matched with a green, even if it can be matched at all. Measures of the luminosity at four or five positions in the spectrum, extending from near the extreme red to the blue, will give an unerring criterion of the kind and extent of colour blindness from which they are suffering. The existence of a neutral point in the spectrum is sufficient to indicate that their blindness is of a nature to be dangerous in certain occupations. To some it may be a difficulty how a neutral point can be found in such cases, since all sensations are more or less present. The reason, however, was explained on page 96.

CHAPTER XV.

EXAMPLES of colour blindness have been brought to your notice, and various measurements made by persons possessing normal and defective colour vision have been recorded, but no attempt has been made to discuss the two leading rival theories that have been laid before you. Regarding these theories you may expect me to say something, and to avow myself a partisan of one or the other. This last I must decline to do, though it will have been seen by the line that I have taken in these lectures that the Young theory attracts me. There are, however, difficulties in adapting it to explain several facts of colour vision which seem to render it, to say the least, incomplete. For instance, to explain the colours produced by simultaneous contrast, the Young theory has to betake itself into psychological ground. I will show you some excellent examples of contrast colours. We have upon the screen a patch of white reflected light, superposed over a patch of red light. Placing a thin rod in the paths of the two beams, we have two shadows—one illuminated by white and the other by red, and lying between them a mixed light of red and white. The shadow illuminated by the white does not appear white, but a bluish-grey. When the spectrum colour is changed to orange the blue is intensified, whilst when it is green, what should be white appears of an orange-salmon colour. Other colours give the white different hues which I need not describe.

These contrast colours are usually said to be *complementary* to the spectrum colours employed, though it must be recollected that what a complementary colour should be is determined by the quality of the white light which the two, when mixed, are made to match. But recent measures of my own show that they are not truly complementary in most instances, whatever the white light may be. But whether they are or not does not much matter when the explanation offered by the followers of the Young theory is considered, for it is asserted that such contrast colours have no real existence, but are psychological, or—what this comes to be—simply delusions. If they are not real colours felt by the retina, they have a very good resemblance to them, and the same series of delusions are so persistent and so constant for all normal vision that they can always be measured as having a constant value. I bear in mind the experiment in which the contrast colour, after being produced, is isolated in the eye from the colour producing it and the background, and the continuance of the hue produced by the contrast. This *retention* may be psychological, but there are no grounds to my mind for saying that its *production* is due to the same cause, more especially as experiments have been arranged to show that one eye may see a contrast colour, whilst the other may see it of its uncontrasted hue. In this last experiment it can scarcely be conceived that one eye should be subject to delusion, whilst the other was free from it. If, then, we may presume that they are real colours, the Young theory fails to explain them, and the explanation offered by the Hering theory is much more acceptable, as it propounds the idea that the retina has to be considered as a whole, and that if (say) red light is at work at one part its complementary colour (blue-green) must be felt at another. It would be still more acceptable had it happened that the contrast colours were truly complementary, and if the same action was noticeable when the adjacent part of the retina was not also stimulated.

For what I may call the straightforward part of

colour vision, dealing with ordinarily bright colours, the Young theory is amply sufficient; but when we come to the feeble luminosities and the colour fields, it is again difficult to adapt to explain the phenomena observed. When we reduce the luminosity of a coloured ray sufficiently we feel the sensation of grey light : no colour is felt. Why is this? On the Hering theory it is capable of the explanation that we have the white sensation left unextinguished, but I fail to see any explanation on the Young theory. When we take colour fields with pure colours (see appendix, page 208), we are met with the unexplained difficulty that the colour from a bright spot of light vanishes almost suddenly towards the periphery of the retina, and is replaced by a bright *white* light, and that the extent of the field depends on the brightness of the colour. This, perhaps, is the most telling observation which can be recorded against the Young theory as it stands at present. It has this support, however, in the *sequence* of the phenomena observed, viz., when the boundary for the colour which we will suppose to be pure red is being taken (as described at page 11), that close to the point where it bursts into pure white, it assumes a pink colour (*i.e.*, a mixture of red and white), whilst, if the red be scarlet, containing according to this theory a little green sensation, it becomes orange before white, showing that the red sensation is dimmed slightly before the green, and so with the other colours. What are called "after images" I have not touched upon so far, nor shall I here, for it is at this point that we step into very debateable ground. The colours perceived in them are, as yet, not capable of being put to the test of physical measurement, and I must leave the psychologist or the physiologist to account for them in their own way.

Viewing the Hering theory from a physical stand-point. and in the light of colour measurement, it appears to be deficient in several respects. To take one point. We have seen that when blue and yellow are mixed together to make white the sum of the luminosities of the two colours separately is equal to the luminosity of the white produced. According to the Hering theory, the yellow colour contains a certain amount of the white-black sensation besides the yellow sensation, as does also the blue colour besides the blue sensation. The theory tells us that when white is produced by the mixture, the blue sensation undoes the work that the yellow sensation has done, and the white sensation is alone left behind. If this be the case, the sum of the separate luminosities cannot be the same as that of the white produced, but should be greater. The theory also has to be strained sometimes to make it fit in with other observed facts. Take, for instance, the case of persons who are called red-blind and green-blind on the Young theory. We are told by the Hering theory that both are red-green-blind—that is, blind to both green and red, and only see blue and yellow—and that the only difference between them is that the former has his spectrum slightly shortened at the red end, the maxima of the yellow-blue sensations being shifted a little further towards the violet end of the spectrum. The natural question to ask is: Why this shift occurs? Surely it is more rational to adopt a theory which does not require such a supposition? If the sensitive matter acted upon by the yellow-blue rays be always of the same chemical composition, the shift cannot occur. It might, perhaps, be allowed that one shift was practicable, but, unfortunately, the shifts must become numerous when the cases of partial colour blindness are to be accounted for, and this would necessitate a constantly varying chemical composition of this matter, and of that acted upon by the red-green rays.

Again, in the extinction of the spectrum, the red and the green sensations in quantities to neutralize one another should be extinguished nearly together, even allowing for what physiologists tell us is the case, that the breaking down, or dissimulation, of cell tissue continues longer than its building up, but we find a large difference between the two. As already indicated, the luminosity curve of the feeble spectrum favours the theory of Hering being that here we only have the white-black sensation, and naturally the persistency curves must be scored in its favour. But the cases of B. C. and M., it seems to me, cannot be explained by the theory without any undue straining or assumptions. If we try and fit the cases of colour blindness due to tobacco scotoma to the theory, we find that in many cases yellow is not recognised, though blue is invariably. If the blue be active, the yellow should also be so.

And here I may remark that it has been assumed that the two classes of colour blindness are due to different causes. A question to ask ourselves is whether all colour blindness may not have been caused originally by disease. In the congenital form, it is true, no disease of the retina is traceable in the eye, and it is usually hereditary, but it does not follow that the want of response of the perceiving apparatus to certain sensations may not have been due to what, for want of a better expression, I may call an hereditary partial paralysis of the perceiving apparatus. If this be so, we have a connecting link between the two classes, and then a perfect theory should explain both classes on the same grounds. The suspicion that the monochromatic vision of P. and Q. might possibly be due to disease before birth, owing to the behaviour of their eyes under certain conditions, would then be explicable. I have no desire to press this view, though it seems to me to be one which is not out of all reason, taking analogies from other defects which are hereditary.

It has been usually accepted that the fields for blue and yellow in the eye are approximately the same, as are those of the green and red, and this has been taken as showing the interdependence between the two pairs according to the Hering theory. It has already been pointed out that the question of extent of fields requires still further investigation beyond that which it has received, and measures made by the method given on page 208 seem to cast a doubt as to whether this interdependence can be upheld. It will be noticed that the fields do not extend proportionately on the nasal and temporal sides (see also Fig. 3). It should also be remarked that the order of extent of field for the different colours does not follow the same order as their disappearance. A point that is sometimes raised in favour of Hering's theory is the negative image formed after the eye is fatigued by looking at bright red or bright green. The negative images (see page 30) are said to be the complementary of these colours. The Young theory tells us that the red or the green sensation suffers fatigue by one or other colour, and that when the eye subsequently rests on a grey surface the other two sensations are chiefly stimulated and cause the complementary colour. It is said that it is easier to produce a *negative green* image than a *negative red* image, and the adherents of Hering tell us that this is due to the fact that destructive action is more readily carried out than constructive. In the Young theory, it is held that the green sensation is always mixed with white, whilst the red is fairly pure, and thus, for equal luminosities, the surplus green sensation is much less stimulated than the red, which offers a consistent explanation of this fact. There are several other minor difficulties in the way of accepting Hering's theory as it stands from a physical point of view, but we need not discuss them now.

The final sensation curves for the spectrum colours on the Young theory are still under consideration, and are not definitely fixed, though the observations made have been very numerous. Recently Helmholtz, in the last edition of his " Physiological Optics," has calculated, from Kœnig's observations, that no one of the three sensations is singly stimulated by any colour, even at the extreme ends of the spectrum, and he makes the three fundamental sensations vary considerably from those given in these pages. Every colour he states is considerably mixed with white light. The calculations by which he arrived at this conclusion are of a complicated nature, and I think if he had had besides the colour equations of Kœnig, the luminosities and the extinction measures before him, there might have been a modification of his views, for these last give evidence to the contrary.

There is a possible modification of the Young theory which would account for a good many of the phenomena that are unaccounted for by it in its present form, though it may raise new difficulties in the minds of some. Let us suppose that each of the three sensations were compounded of fundamental *light and of colour* in fixed and definite proportions, and not in the same proportion in each; and further that the apparatus in the eye which was responsible for each sensation had two functions, one of which was to respond to the fundamental light sensation and the other to the colour. One essential difference between this modification of the Young theory and that of Hering is that, whilst in the latter the white sensation is a sensation *distinct from the colour sensations*, in the former it is a *definite part* of them. The fact that the sensation of colour is lost before the sensation of light is one of the greatest significance, and any theory to be accepted must offer a reasonable explanation of it. If the modification suggested be made, it accounts for the existence of this residuum of light equally as well as Hering's theory, and without its drawback. It is not hard to imagine the apparatus which gives rise to two sensations, on the assumption of different kinds of atomic motion, induced by the ether motion, or at least three kinds are possible. When extinction of *colour* is made, the ether vibrations would have sufficient energy to induce but one kind of motion; and when all *light* was extinguished from the same ray, they would not be capable of inducing any sensible motion whatever. In the case of Miss W., who saw all colours as white, it might be that disease had entirely prevented the first kind of motion in all three sensations, and that in P. and Q. the red and green sensations were absent or paralysed in their entirety, whilst the blue sensation was left in full operation. In B. C. the blue and red sensations would be similarly absent, leaving the green sensation unchanged. The coincidence of their persistency and luminosity curves would then indicate that the proportions of fundamental light and colour remained the same throughout. Other examples and considerations seem to indicate that the proportion of colour to fundamental light is greatest in the red sensation, next in the green, and least in the blue. This would explain why with increasing intensities blue appears white sooner than green, and much sooner than red. The proposed modification would also offer the necessary explanation as to the disappearance of colour from the field.

Looking at colour vision from what I may call an evolutionary point of view, the " light-colour " theory commends itself as probable. There are many reasons for thinking that the visual sensation first evolved was that of light, subsequently followed by that of colour. The first evolved colour sensation would appear to have been the blue, and the last the red. The discussion of this hypothesis would carry me beyond my limits, and I must leave it thus baldly expressed for your consideration.

For my own part, whatever theory of colour sensations may prove to be the right one, I lean strongly to the idea that the cause of vision will be found in chemical

action, induced by the impact of the different wave-
lengths of light falling on sensitive matter. A white
substance may absorb all the wave-lengths found in the
spectrum, and if it have three sets of molecules, one of
which has an atom or atoms vibrating with the same
period as the waves of light which show a maximum
for one sensation and another for another, and so on, the
requirements for the colour sensations are met. It may
be that the sensitive part of the retina is like a photographic plate, but with this essential difference—that the
sensitive material is constantly changing. A photographic plate receives an impression which is not
recognisable by the eye, though it can be shown that a
change in the material does take place during the
impact of light, by electrical and other means. When
the eye receives an impression of light, Dewar has
shown that in this case also a current of electricity is
generated. Recent published experiments of my own
have demonstrated that with a low intensity of light,
the chemical change that occurs in a photographic salt
is by no means proportionate to that which takes place
with a greater intensity. In the eye, too, there is a
limit of sensibility to very feeble light. Again, the
curves of the stimulation of the colour sensations to the
spectrum are closely of the same form as the curves

of sensitiveness of the various sensitive salts used by photographers. These are analogies and, of course, must not be pressed too far. There must be such a complexity in the sensitive material in the eye, both chemical and physiological, that it may be that the changes induced by light on the sensitive surface of the retina have to be considered from both aspects. The purely chemical change is naturally that to which a physicist is most prone to incline, and his bias must be discounted, as must also that of the physiologist.

APPENDIX.

THE following is extracted from Maxwell's paper.

The following table contains the means of four sets of observations by the same observer (K.) :—

TABLE IV. (K.)

$$44 \cdot 3 \ (20) + 31 \cdot 0 \ (44) + 27 \cdot 7 \ (68) = W.$$
$$16 \cdot 1 \ (28) + 25 \cdot 6 \ (44) + 30 \cdot 6 \ (68) = W.$$
$$22 \cdot 0 \ (32) + 12 \cdot 1 \ (44) + 30 \cdot 6 \ (68) = W.$$
$$6 \cdot 4 \ (24) + 25 \cdot 2 \ (36) + 31 \cdot 3 \ (68) = W.$$
$$15 \cdot 3 \ (24) + 26 \cdot 0 \ (40) + 30 \cdot 7 \ (68) = W.$$
$$19 \cdot 8 \ (24) + 35 \cdot 0 \ (46) + 30 \cdot 2 \ (68) = W.$$
$$21 \cdot 2 \ (24) + 41 \cdot 4 \ (48) + 27 \cdot 0 \ (68) = W.$$
$$22 \cdot 0 \ (24) + 62 \cdot 0 \ (52) + 13 \cdot 0 \ (68) = W.$$
$$21 \cdot 7 \ (24) + 10 \cdot 4 \ (44) + 61 \cdot 7 \ (56) = W.$$
$$20 \cdot 5 \ (24) + 23 \cdot 7 \ (44) + 40 \cdot 5 \ (60) = W.$$
$$19 \cdot 7 \ (24) + 30 \cdot 3 \ (44) + 33 \cdot 7 \ (64) = W.$$
$$18 \cdot 0 \ (24) + 31 \cdot 2 \ (44) + 32 \cdot 3 \ (72) = W.$$
$$17 \cdot 5 \ (24) + 30 \cdot 7 \ (44) + 44 \cdot 0 \ (76) = W.$$
$$18 \cdot 3 \ (24) + 33 \cdot 2 \ (44) + 63 \cdot 7 \ (80) = W.$$

X.—REDUCTION OF THE OBSERVATIONS.

By eliminating W from the equations above by means of the standard equation, we obtain equations involving each of the fourteen selected colours of the spectrum, along with the three standard colours; and by transposing the selected colour to one side of the equation, we obtain its value in terms of the three standards. If any of the terms of these equations are negative, the equation has no physical interpretation as it stands; but by transposing the negative term to the other side it becomes positive, and then the equation may be verified.

The following table contains the values of the fourteen selected tints in terms of the standards. To avoid repetition, the symbols of the standard colours are placed at the head of each column: —

TABLE VI.

Observer (K.)	(24)	(44)	(68)
44·3 (20) =	18·6	+ 0·4	+ 2·8
16·1 (28) =	18·6	+ 5·8	− 0·1
22·0 (32) =	18·6	+ 19·3	− 0·1
25·2 (36) =	12·2	+ 31·4	− 0·8
26·0 (40) =	3·3	+ 31·4	− 0·2
35·0 (46) =	− 1·2	+ 31·4	+ 0·3
41·4 (48) =	− 2·6	+ 31·4	+ 3·5
62·0 (52) =	− 3·4	+ 31·4	+ 17·5
61·7 (56) =	− 3·1	+ 21·0	+ 30·5
40·5 (60) =	− 1·9	+ 7·7	+ 30·5
33·7 (64) =	− 1·1	+ 1·1	+ 30·5
32·3 (72) =	+ 0·6	+ 0·2	+ 30·5
44·0 (76) =	+ 1·1	+ 0·7	+ 30·5
63·7 (80) =	+ 0·3	− 1·8	+ 30·5

Mr. James Simpson, formerly student of Natural Philosophy in my class, has furnished me with thirty-three observations taken in good sunlight. Ten of these were between the two standard colours, and give the following result :—

$$33 \cdot 7 \ (88) + 33 \cdot 1 \ (68) = W.$$

The mean errors of these observations were as follows :—

Error of $(88) = 2 \cdot 5$; of $(68) = 2 \cdot 3$; of $(88) + (68)$
$= 4 \cdot 8$; of $(88) - (68) = 1 \cdot 3$.

The fact that the mean error of the sum was so much greater than the mean error of the difference, indicates that in this case, as in all others that I have examined, observations of equality of tint can be depended on much more than observations of equality of illumination or brightness.

From six observations of my own, made at the same time, I have deduced the " trichromic" equation—

$$22 \cdot 6 \ (104) + 26 \ (88) + 37 \cdot 4 \ (68) = W \ . \ . \ . \ (2)$$

If we suppose that the light which reached the organ of vision was the same in both cases, we may combine these equations by subtraction, and so find

$$22 \cdot 6 \ (104) - 7 \cdot 7 \ (88) + 4 \cdot 3 \ (68) = D. \ . \ . \ . \ (3)$$

where D is that colour, the absence of the sensation of which constitutes the defect of the dichromic eye.

The sensation which I have in addition to those of the dichromic eye is therefore similar to the full red (104), but different from it in that the red (104) has 7·7 of green (88) in it which must be removed, and 4·3 of blue (68) substituted. This agrees pretty well with the colour which Mr. Pole* describes as neutral to him, though crimson to others. It must be remembered, however, that different persons of ordinary vision require different proportions of the standard colours, probably owing to differences in the absorptive powers of the media of the eye, and that the above equation (2), if observed by K., would have been

$$23 (104) + 32 (88) + 31 (68) = W \ . \ . \ . \ . \ . \ (4)$$

and the value of D, as deduced from these observers, would have been

$$23 (104) - 1·7 (88) - 1·1 (68) = D \ . \ . \ . \ . \ (5)$$

in which the defective sensation is much nearer to the red of the spectrum. It is probably a colour to which the extreme red of the spectrum tends, and which differs from the extreme red only in not containing that small proportion of " yellow " light which renders it visible to the colour blind.

* Philosophical Transactions, 1859, Part I., p. 329.

From other observations by Mr. Simpson the following results have been deduced :—

TABLE A.

	(88)	(68)		(88)	(68)
(99·2 +) =	33·7	1·9	100 (96) =	108	7
31·3 (96) =	33·7	2·1	100 (92) =	120	5
28 (92) =	33·7	1·4	100 (88) =	100	0
33·7 (88) =	33·7	0	100 (84) =	61	11
54·7 (84) =	33·7	6·1	100 (82) =	47	21
71 (82) =	33·7	15·1	100 (80) =	34	33
99 (80) =	33·7	33·1	100 (78) =	22	47
70 (78) =	15·7	33·1	100 (76) =	10	59
56 (76) =	5·7	33·1	100 (72) =	1	92
36 (72) =	0·3	33·1	100 (68) =	0	100
33·1 (68) =	0	33·1	100 (64) =	0	83
40 (64) =	0·2	33·1	100 (60) =	3	60
55·5 (60) =	1·7	33·1			
57 −) =	0·3	33·1			

In the table on the left side (99·2 +) means the whole of the spectrum beyond (99·2) on the scale, and (57—) means the whole beyond (57) on the scale. The position of the fixed lines with reference to the scale was as follows :—

A, 116 ; a, 112 ; B, 110 ; C, 106 ; D, 98·3 ; E, 88 ;
F, 79 ; G, 61 ; H, 44.

The values of the standard colours in different parts of the spectrum are given on the right side of the above table, and are represented by the curves of

Fig. 9, Plate II., where the left-hand curve represents
the intensity of the " yellow" element, and the right-
hand curve that of the " blue " element of colour as it
appears to the colour blind.

The appearance of the spectrum to the colour blind
is as follows :—

From A to E the colour is pure " yellow," very faint
up to D, and reaching a maximum between D and E.
From E to one-third beyond F towards G the colour is
mixed, varying from " yellow " to " blue," and becoming
neutral or " white " at a point near F.   In this part of
the spectrum the total intensity, as given by the dotted
line, is decidedly less than on either side of it, and near
the line F, the retina close to the " yellow spot " is less
sensible to light than the parts further from the axis
of the eye.   This peculiarity of the light near F is
even more marked in the colour blind than in the
ordinary eye.   Beyond F the " blue" element comes to
a maximum between F and G, and then diminishes
towards H, the spectrum from this maximum to the
end being pure " blue."

The results given above were all obtained with the
light of white paper, placed in clear sunshine.   I have
obtained similar results when the sun was hidden, by
using the light of uniformly illuminated clouds, but I

do not consider these observations sufficiently free from disturbing circumstances to be employed in calculation. It is easy, however, by means of such observations, to verify the most remarkable phenomena of colour blindness, as, for instance, that the colours from red to green appear to differ only in brightness, and that the brightness may be made identical by changing the width of the slit ; that the colour near F is a neutral tint, and that the eye in viewing it sees a dark spot in the direction of the axis of vision ; that the colours beyond are all blue of different intensities, and that any " blue " may be combined with any " yellow " in such proportions as to form " white." These results I have verified by the observations of another colour-blind gentleman, who did not obtain sunlight for his observations ; and as I have now the means of carrying the requisite apparatus easily, I hope to meet with other colour-blind observers, and to obtain their observations under more favourable circumstances.

Measurements of Colour Fields.

Some experiments in the measurement of the colour fields in the horizontal direction with the pure spectrum colours will help to show what importance is to be attached to the luminosity of the colour and the size of the spot of light with which the observations are made. A yellow and a blue of the spectrum were taken of such hues that when mixed they formed a patch of white light similar to the electric light. Their luminosities were measured, and the yellow found to be $1\cdot6$ of the light of an amyl-acetate lamp or $1\cdot28$ standard candles; the blue was $\frac{1}{24}$ of this luminosity. The fields for these two colours were measured by automatically throwing spots of each colour separately on a white card which moved round a centre over which the eye was placed. The light was subsequently diminished to $\frac{1}{2}$, $\frac{1}{4}$, and $\frac{1}{8}$ of the above values, and readings again made. The following results were obtained with a spot of $\cdot7$ inch diameter :—

Light.	Yellow.		Blue.	
	Nasal side.	Temporal side.	Nasal side.	Temporal side.
Full	33°	45°	35°	45°
$\frac{1}{2}$	24°	36°	26°	38°
$\frac{1}{4}$	18°	24°	22°	32°
$\frac{1}{8}$	11°	15°	19°	30°

With a spot of $\cdot3$ inch diameter the following were obtained :—

Full	24°	32°	21°	27°
$\frac{1}{2}$	17°	28°	16°	22°
$\frac{1}{4}$	13°	16°	14°	20°
$\frac{1}{8}$	8°	10°	13°	16°

It will be evident how the field contracts as the light is diminished in brightness, and also that the blue field does not diminish equally with the yellow field, but is more persistent. Again, it will be noticed that the luminosity of the blue, for the same extent of field to be covered, has to be much lower than for the yellow.

The diminished area of the spot of light also diminishes the field, and the same order of diminution of field is obtained as with the larger spot.

Another set of experiments, made with the same aperture of slit passed through the spectrum, and the field taken at different points, give the following results :—

Spectrum scale. (See Fig. 41, page 210.)	Nasal side.	Temporal side.
58·6	18°	35°
54·6	27°	46°
50·6	33°	47°
46·6	25°	30°
42·6	21°	21°
38·6	17°	17°
34·6	22°	30°
30·6	25°	33°
26·6	33°	40°
22·6	37°	44°
18·6	28°	40°
14·6	22°	34°
8·6	20°	30°

P

Here we see that although the luminosity of the colour spots varies at the spectrum luminosity, the

FIG. 41.

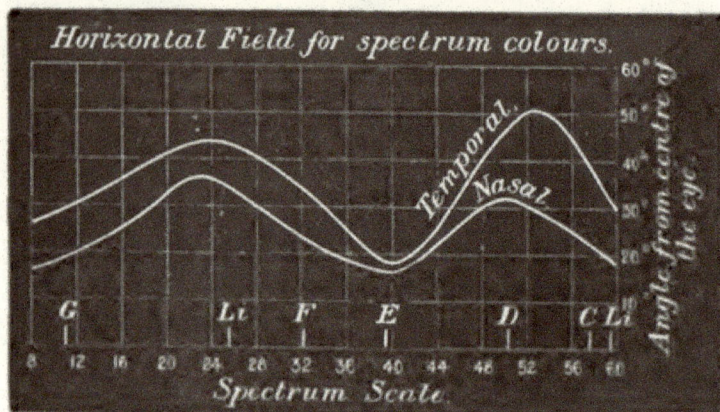

Horizontal Field for spectrum colours.

fields do not vary proportionally; when the luminosities of the green, yellow and red are made equal, the fields become nearly equal on the nasal side. The field for the blue, however, then becomes vastly larger than that for the others, showing a peculiarity which is very remarkable.

Recently published experiments on colour fields have been so largely based on the exigencies of the Hering theory, that it is somewhat difficult to decide their significance from any other aspect.

TABLE I.—LUMINOSITY CURVES FOR THE NORMAL EYE (see Fig. 20).

I. Scale number.	II. Wave-length.	III. Outside yellow spot.	IV. Yellow spot.	V. Fovea centralis.
64	7217			
63	7082	..	1	
62	6957	1	2	2
61	6839	2	4	4
60	6728	3·5	7	8
59	6621	7·5	12·5	15·5
58	6520	12·5	21	24
57	6423	19	33	37·5
56	6330	27·5	50	60
55	6242	35	65	77
54	6152	43	80	90
53	6074	52·5	90	97
52	5996	61·0	96	100
51	5919	71·0	99	100
50	5850	79·0	100	98
49	5783	84	99	95
48	5720	85	97	90
47	5658	83·5	92·5	85
46	5596	81·0	87	79
45	5538	77·0	81	72·5
44	5481	72.5	75	66
43	5427	68·0	69	59
42	5373	62.5	62·5	51
41	5321	57	57	45

TABLE I.—*continued.*

I.	II.	III.	IV.	V.
Scale number.	Wave-length.	Outside yellow spot.	Yellow spot.	Fovea centralis.
40	5270	52	50	40
39	5221	46	42·5	32
38	5172	41·5	36	27·5
37	5128	37·5	29·5	22·0
36	5085	33·5	24	18
35	5043	30·0	18·2	14
34	5002	26·5	14·2	10
33	4963	24	10·5	8·4
32	4924	21	8·5	6·5
31	4885	18·5	7·0	5·5
30	4848	16·5	5·5	4·0
29	4812	14·5	4·7	3·5
28	4776	13·0	4·0	3·0
27	4742	11·5	3·5	2·0
26	4707	10.5	2·8	2·4
25	4675	9·4	2·3	2·1
24	4639	8·2	1·82	1·9
23	4608	7·3	1·6	1·5
22	4578	6·3	1·4	
21	4548	5·7	1·2	
20	4517	5·0	1.08	1·0
19	4488	4·5	·94	
18	4459	4·0	·86	
17	4437	3·6	·78	
16	4404	3·1	·70	

TABLE I.—*continued.*

I.	II.	III.	IV.	V.
Scale number.	Wave-length.	Outside yellow spot.	Yellow spot.	Fovea centralis.
15	4377	2·7	·62	·62
14	4349	2·3	·56	
13	4323	2·1	·50	
12	4296	1·9	·45	
11	4271	1·65	·40	
10	4245	1·4	·34	
9	4221	1·2	·30	
8	4197	1·0	·26	
7	4174	·88	·22	
6	4151	·75	·18	
5	4131	·63	·16	
4	4106	·50	·14	

Tables II. and III.—Curves of Luminosity of a Partially
Red-Blind and of a Partially Green-Blind Person (see
Fig. 23).

Scale number.	Wave-length.	Luminosity.	
		Red-blind.	Green-blind.
64	7217	0	1
62	6957	1	2
60	6728	2	7
58	6520	6	21
56	6330	12	50
54	6152	26	80
52	5996	49	96
50	5850	70	98
48	5720	77	93
46	5596	77	83
44	5481	70	70
42	5373	61	55
40	5270	47	40
38	5172	34	27
36	5085	23	18
34	5002	14	10
32	4924	8·5	5·5
30	4848	5·5	3·0
28	4776	4·0	2·5
26	4707	2·7	2·0
24	4639	1·8	1·8
22	4578	1·35	1·4
20	4517	1·1	1·1

TABLE IV.—LUMINOSITY OF SPECTRUM REDUCED IN INTENSITY, SO
THAT $D = \frac{1}{12\cdot6}$ AMYL LAMP 1 FOOT DISTANT (see Fig. 25).

Scale number.	Mean reading.	Mean reading, reduced to 100 max.	P. and Q.'s readings, 100 max.	Persistency curve for the centre of the eye.
55·6	·5	·6	2	2
53·6	5·5	7·0	3·6	3·6
51·6	13	16·7	8	8
49·6	23	29·7	22	22
47·6	40	50·0	44	44
45·6	57	71·2	69	69
43·6	70	87·5	93	93
41·6	79	98·7	100	99·5
39·6	78	97·5	99·5	98·5
37·6	74	92·5	96	93
35·6	66	82·5	89	84
33·6	55	68·7	77·5	71
31·6	44·5	55·2	61	53·5
29·6	35	43·7	45·5	36·5
27·6	24	30·0	33·5	24
25·6	17	21·7	25	16
23·6	13	16·7	18	10
21·6	10	12·5	13	8
19·6	8	10·0	9·5	6
13·6	3	3·7	4·2	3
9·6	2	2·5	2·5	2

TABLE V.—LIMIT OF COLOUR VISION (see Fig. 26.)

Scale Number.	Wave-Length.	Mean reading of the colour limit of the spectrum D, being 1 amyl lamp in $\frac{1}{10000}$ths.	Luminosity of the ordinary spectrum.	Luminosity of the rays when each colour disappears, each ray having the original luminosity of 1 amyl lamp in $\frac{1}{10000}$ths.
61	6839	120	4	48·0
60	6728	67	7	46·9
58	6520	26	21	54·6
56	6330	13	50	65·0
54	6152	9·5	80	76·0
52	5996	9·0	96	86·4
50	5850	9·0	100	90·0
48	5720	9·0	97	87·3
44	5481	9·5	75	71·3
40	5270	10·5	50	52·5
36	5085	12·5	24	30·0
32	4924	18	8·5	15·3
28	4776	32	4·0	12·8
24	4639	55	1·8	12·0
20	4517	90	1·08	9·7
16	4404	160	·70	11·2
12	4296	250	·45	11·0
8	4197	400	·26	10·4
4	4106	700	·14	9·8

TABLE VI.—EXTINCTION BY CENTRAL PORTION OF NORMAL EYE
(see Fig. 28).

I.	II.	III.	IV.	V.	VI.
Scale number.	Wave-length.	E. Reduction of original luminosity in millionths to cause extinction.	L. Lumino-sity of original beam.	$\dfrac{E \times L}{100}$.	Persistency curve $\dfrac{650}{E}$ (Maximum = 100).
64	7217	55,000			
63	7082	30,000	1	300·0	
62	7957	15,000	2	300·0	
61	6839	7500	4	300·0	
60	6728	3750	7	262·5	
59	6621	1900	12·5	237·5	·34
58	6520	1050	21	220·5	·62
57	6423	650	33	214·5	1·0
56	6333	380	50	190·0	1.71
55	6242	272	65	176·8	2·38
54	6152	196	80	156·0	3·32
53	6074	140	90	126·0	4·64
52	5996	97	96	93·12	6·70
51	5919	57	99	56·43	11·40
50	5850	35	100	35·0	18·6
49	5783	24	99	23·76	27·1
48	5720	17	97	16.49	38·2
47	5658	12·6	92·5	11·65	51·6
46	5596	10·2	87	8·87	63·7
45	5538	8·6	81	6·97	75·6
44	5481	7·4	75	5·55	87·8

TABLE VI.—*continued.*

I.	II.	III.	IV.	V.	VI.
Scale number.	Wave-length.	E. Reduction of original luminosity in millionths to cause extinction.	L. Luminosity of original beam.	$\frac{E \times L}{100}$.	Persistency curve $\frac{650}{E}$ (Maximum = 100).
43	5427	6·7	69	4·62	97·0
42	5373	6·55	62·5	4·09	99·5
41	5321	6·5	57	3·705	100
40	5270	6·55	50	3·27	98·5
39	5221	6·65	42·5	2.83	97·5
38	5172	6·85	36	2·46	95·0
37	5128	7·2	29·5	2·12	90·0
36	5085	7·6	24	1·82	81·3
35	5043	8·15	18·2	1·48	80·0
34	5002	8·8	14·2	1·25	74·0
33	4963	10·2	10·5	1·07	63·0
32	4924	11·6	8·5	·988	56·0
31	4885	13·6	7·0	·952	47·7
30	4848	16·3	5·5	·896	40·0
29	4812	20·5	4·7	·963	31·7
28	4776	26·0	4·0	1·040	25·0
27	4742	31·0	3·5	1·085	20·9
26	4707	38·5	2·8	1·078	16·9
25	4674	46·0	2·3	1·058	14·1
24	4639	56·0	1·82	1·019	11·6
23	4608	67·0	1·6	1·072	9·7
22	4578	80	1·4	1·120	8·41

TABLE VI.—*continued.*

I.	II.	III.	IV.	V.	VI.
Scale number.	Wave-length.	E. Reduction of original luminosity in millionths to cause extinction.	L. Lumino-sity of original beam.	E × L / 100.	Persistency curve 650 / E (Maximum = 100).
21	4548	95	1·2	1·140	7·22
20	4517	107	1·08	1·156	6·1
19	4488	124	·94	1·165	5·23
18	4459	140	·86	1·204	4·64
17	4437	160	·78	1·228	4·1
16	4404	180	·70	1·260	3·60
15	4377	200	·62	1·240	3·25
14	4349	220	·56	1·232	2·95
13	4323	240	·50	1·200	2·7
12	4296	270	·45	1·215	2·4
11	4271	300	·40	1·200	2·18
10	4245	335	·34	1·139	1·94
9	4221	375	·30	1·125	1·73
8	4197	430	·26	1·118	1·51
7	4174	490	·22	1·078	1·32
6	4151	510	·18	·918	1·27
5	4131	640	·16	1·024	1·01
4	4106	750	·14	1·050	0·86

TABLE VII.—EXTINCTION BY WHOLE EYE
(see Fig. 28).

I.	II.	III.	IV.	V.	VI.
Scale number.	Wave-length.	E. Reduction of original luminosity in millionths to cause extinction.	L. Luminosity of original beam.	$\frac{E \times L}{100}$.	Persistency curve $\frac{650}{E}$ (Maximum= 100).
38	5172	6·9	41·5	2·86	94·2
37	5128	7·1	37·5	2·66	91·6
36	5085	7·4	33·5	2·48	87·8
35	5043	7·7	30·0	2·31	84·4
34	5002	8·0	26·5	2·12	81·2
33	4963	8·4	24·0	2·02	77·5
32	4924	8·8	21·0	1·85	73·8
31	4885	9·4	18·5	1·74	69·2
30	4848	10·0	16·5	1·65	65·0
29	4812	10·7	14·5	1·55	60·6
28	4776	11·5	13·0	1·49	56·5
27	4742	13·0	11·5	1·49	50·0
26	4707	14·5	10·5	1·52	44·8
24	4639	18·5	8·2	1·52	34·1
22	4578	23·0	6·3	1·45	28·3
20	4517	30·0	5·0	1·50	21·7
18	4459	39·0	4·0	1·56	16·7
16	4404	51	3·1	1·59	12·3

TABLE VII.—*continued.*

I.	II.	III.	IV.	V.	VI.
Scale number.	Wavelength.	E. Reduction of original luminosity in millionths to cause extinction.	L. Luminosity of original beam.	$\dfrac{E \times L}{100}$	Persistency curve $\dfrac{650}{E}$ (Maximum = 100).
14	4349	66	2·3	1·52	9·85
12	4296	80	1·9	1·52	8·12
10	4245	110	1·4	1·54	5·91
8	4197	154	1·0	1·54	4·22
6	4151	204	·75	1·54	3·18
4	4106	307	·5	1·54	2·11
2	4063	513	·3	1·54	1·26
0	4020	770	·2	1·54	·84

From 38 to 64 the extinction is the same as with the central part of the eye.

TABLE VIII.—P.'s CURVES * (see Fig. 31).

I.	II.	III.	IV.	V.	VI.	VII.
Scale number.	Wave-length.	Mean reading of extinction in millionths of original luminosity.	Adopted reading in millionths of original luminosity.	Persistency curve 680/ad. reading	P.'s luminosity curve.	Absolute luminosity of extinction. IV. × VI. / 14
52	5996	68	68	10	7	34
50	5850	35	35	19·4	19	47·5
48	5720	17	17	40	39	47·3
46	5596	10·2	10	68	65	46·4
45	5538	9·3	9·0	76	76	48·8
44	5481	8·0	8·1	84	90	52·8
42	5373	7·2	7·2	94·5	98	50·3
40	5270	6·7	6·8	100	99	48·1
38	5172	7·2	7·0	97	97·5	48·7
36	5085	8·05	7·7	90	90	49·5
34	5002	8·05	8·4	81	80	47·9
32	4924	9·9	9·8	69	65	45·5
30	4848	13·2	12·5	54	50	44·6
28	4776	13·9	15·0	45·3	36	38·6
27	4742	16·8	17·0	40	31·5	38·2
26	4707	21·6	20·5	32	26·5	38·8
24	4639	30	27	25	19·5	37·6
22	4578	36	35	19	14	35
20	4517	42	45	15·5	10	32·2
16	4404	79	79	8·5	5·5	31·2
10	4245	180	190	3·6	2·5	32·2
6	4151	270	270	2·7		

* In this and the next two Tables the intensity of the illumination of the D ray before reduction is equal to that of an amyl-acetate lamp at one foot from a screen. The figures in Col. VII. are in millionths of the illumination of an amyl-acetate lamp at one foot distant, every ray being made of that intensity.

Table IX.—H. R.'s Curves (see Fig. 32).

I.	II.	III.	IV.	V.	VI.	VII.
Scale number.	Wave-length.	Mean reading of extinction in millionths of original luminosity.	Adopted reading in millionths of original luminosity.	Persistency curve 590 ad. reading	Luminosity curve.	Absolute luminosity of extinction IV. × VI. / 48
57	6423	1200	1200	·49	5	125
56	6330	900	850	·69	7	124
55	6242	500	550	1·07	10	115
54	6152	250	250	2·36	17	88
53	6074	..	150	3·93	25	78
52	5996	90	90	6·56	35	66
51	5919	60	45	13·1	47	44
50	5850	27	27	21·8	57	32
48	5720	18	15	39·3	66	21
46	5596	10	10	59	69	14
44	5481	9·3	8	73·8	64	11
42	5373	6·5	6·2	95·1	56·5	7
40	5270	5·9	5·9	100	45	5·5
38	5172	6	6	98·3	32	4
36	5085	..	6·6	89·4	20	2·7
35	5043	7	7·2	81·9	16	2·4
34	5002	..	8	73·8	12·5	2·1
32	4924	10	9·6	61·5	8	1·6
30	4848	11·5	12	49·2	6	1·5
28	4776	14·5	14·5	40·7	5	1·5
26	4707	20	17·5	33·7	4	1·5
24	4639	20	22	26·8	3	1·4
22	4578	..	30	19·7	2·4	1·5
18	4459	55	57	10·4	1·3	1·5
14	4349	115	115	5·1	·7	1·7
10	4245	..	160	3·7	·5	1·7
6	4151	200	200	2·9	·4	1·7

Table X.—V. H.'s Curves (see Fig. 33).

I. Scale number.	II. Wavelength.	III. Mean reading of extinction in millionths of original luminosity.	IV. Adopted reading in millionths of original luminosity.	V. Persistency curve 530 ad. reading	VI. Luminosity curve.	VII. Absolute luminosity of extinction $\frac{IV. \times VI.}{75}$.
57	6423	500	500	1·1	31	206
56	6330	350	350	1·5	43	200
54	6152	200	180	2·9	61	146·4
52	5996	100	100	5·3	70	93·3
50	5850	40	40	13·3	73	38·9
48	5720	..	25	21·2	69	23
46	5596	10	10	53·0	63	8·4
45	5538	6·5	6·5	81·6	58	5·0
44	5481	6·0	5·7	93	54	4·1
42	5373	5·5	5·3	100	46	3·3
40	5270	5·5	5·4	98·2	36	2·6
38	5172	5·7	5·7	93	24	1·8
36	5085	6·7	6·5	81·6	15	1·3
34	5002	7·0	7·0	75·7	9·5	·89
32	4924	8·5	8·5	62·3	7·0	·79
30	4848	10·7	10·5	50·5	5·0	·70
28	4776	16	16	33·1	3·7	·79
26	4707	..	22·5	23·5	2·7	·81
24	4639	30	31	17·1	1·82	·75
22	4578	42·5	42	12·6	1·4	·78
20	4517	55	55	9·6	1·0	·73
16	4404	105	100	5·3	·7	·93
12	4296	175	170	3·1	·45	1·02
10	4245	200	200	2·7	·34	·91

TABLE XI.—B. C.'s CURVES (see Fig. 34).

I.	II.	III.	IV.	V.	VI.
Scale number.	Wave-length.	Adopted reading in hundred thousandths.	Persistency curve 12,500 readings in V.	Lumino-sity of original beam.	Absolute luminosity of extinction III. and V.
61	6839	7500	1·6		
60	6728	5500	2·3	·5	27·5
59	6622	4000	3·1	1	40
58	6520	2800	4·5	2	56
57	6423	2000	6·2	4	80
56	6330	1500	8·3	6	90
55	6242	1150	10·8	8	92
54	6152	950	13·1	11·5	109·2
53	6074	750	16·6	16	120
52	5996	580	21·6	21·5	125
51	5919	450	29	28·5	122·5
50	5850	350	36	37	129·5
49	5783	275	45·5	47	129·2
48	5720	215	58	60	129
47	5658	170	73·4	76	129·2
46	5596	140	89·3	92	129
45	5538	125	100	98	122·5
44	5481	125	100	100	125
43	5427	130	96·1	97	126
42	5373	150	83	85	127·5
41	5321	180	69·4	65	117
40	5270	215	59	45	96·7
39	5221	250	50	30	75

TABLE XI.—*continued.*

I.	II.	III.	IV.	V.	VI.
Scale number.	Wave-length.	Adopted reading in hundred thousandths.	Persistency curve 12,500 readings in V.	Lumino-sity of original beam.	Absolute luminosity of extinction III. and V.
38	5172	290	43	1·5	723.2
37	5128	335	37	16	53·6
36	5055	380	33	11·5	43·7
34	5002	500	25	7	35
32	4994	650	19	4	26
30	4848	850	14	2·5	23·3
28	4776	1100	11·4	2	22
26	4707	1500	8·3	1·5	22
24	4639	2000	6·2	1	20
22	4578	2700	4·6	5	13·5
18	4459	4750			
14	4349	7500			
10	4245	11000			

TABLE XII.—M.'s LUMINOSITY CURVE COMPARED WITH THE NORMAL
(see Fig. 30).

I.	II.	III.	IV.	V.	VI.	VII.
Scale number.	Wave-length.	Mean reading.	Mean reading × 1·8.	Normal luminosity curve, centre of eye.	Difference of last two columns.	Difference × 5·15.
61	6839	2	3·6	4	·4	2·57
59	6621	7	12·6	12·5	—·1	·51
57	6423	18	32·4	33	+·6	3·09
55	6242	36	64·8	65	·2	1·03
53	6074	49	88·2	89·5	1·3	6·71
52	5996	52	95·4	96·5	1·1	5·66
51	5919	54	97·2	99·5	2·3	11·8
50	5850	54	97·2	100	2·8	14·4
49	5782	52·5	94·5	99·5	5·0	25·7
48	5720	50	90	97	7·0	36·0
47	5658	46	82·8	92·5	9·7	49·9
46	5596	41	73·8	87	13·2	68·0
44	5481	32	57·6	75	17·4	89
42	5373	23	43·2	62.5	19·3	99
40	5270	17	30·6	50	19·4	100
38	5172	10	17·5	35·5	18	93
36	5085	4	7·2	24	16·8	86·5
34	5002	1·0	1·8	14·5	12·7	65·5
31	4885	·5	·7	6·5	5·8	37·7
28	4776	0	0	4	4	20·6

TABLE XIII.—MISS W.'s CURVES (see Fig. 39).

Scale number.	Wave-length.	Readings.	Extinction in $\frac{1}{100000}$.	Persistency curve.
63	7082	0		
62	6957	1		
60	6728	7		
58	6520	18		
57	6423	28		
56	6330	43		
54	6152	76	900	2
52	5996	90	250	7
50	5850	95	130	13·5
48	5720	93	60	29
46	5596	83	34	51
44	5481	71	22	80
42	5321	58	18·5	92
40	5270	46	17·5	100
38	5172	32	18	94
36	5085	21	19·5	90
34	5002	12·5	22	79
32	4924	7	27	65
30	4848	4·5	34	51
28	4776	3·0	40	38·5
25	4675	1·5	60	29
20	4518	0·4	250	7
19	4488	0·0	350	5
16	4404	—	600	

INDEX

WORKS ON PHOTOGRAPHY

BY

CAPT. W. DE W. ABNEY, C.B., D.C.L., F.R.S.,

Late Royal Engineers.

INSTRUCTION IN PHOTOGRAPHY. Post 8vo., 3s. 6d.

PHOTOGRAPHY WITH EMULSIONS. Post 8vo., 3s.

NEGATIVE MAKING. Post 8vo., 1s.

ART AND PRACTICE OF SILVER PRINTING. Written in conjunction with Mr. H. P. ROBINSON. Post 8vo , 2s. 6d.

PLATINOTYPE. Written in conjunction with LYONEL CLARKE. Post 8vo., 2s. 6d.

LONDON: SAMPSON LOW, MARSTON & COMPANY, LIMITED,
ST. DUNSTAN'S HOUSE, FETTER LANE, FLEET STREET, E.C.

LONDON: PRINTED BY WILLIAM CLOWES AND SONS, LIMITED,
STAMFORD STREET AND CHARING CROSS.

www.ingramcontent.com/pod-product-compliance
Lightning Source LLC
Chambersburg PA
CBHW021527210326
41599CB00012B/1403